Psychopharmacology Reconsidered

Robert Haim Belmaker • Pesach Lichtenberg

Psychopharmacology Reconsidered

A Concise Guide Exploring the Limits
of Diagnosis and Treatment

 Springer

Robert Haim Belmaker
Division of Psychiatry
Ben-Gurion University of the Negev
Beersheva, Israel

Pesach Lichtenberg
The Hebrew University in Jerusalem
Jerusalem, Israel

ISBN 978-3-031-40370-5   ISBN 978-3-031-40371-2   (eBook)
https://doi.org/10.1007/978-3-031-40371-2

This Springer imprint is published by the registered company Springer Nature Switzerland AG
The registered company address is: Gewerbestrasse 11, 6330 Cham, Switzerland

Paper in this product is recyclable.

This book is dedicated to the memory of Prof. Thomas A Ban who inspired and encouraged the writing of this book and devoted a weekly zoom consultation to the topics of this volume in the last year of his life up until the week before his death at age 92.

Acknowledgements

The authors are indebted to Prof. Yuly Bersudsky, Dr. Alex Clayman, Prof. Haim Knobler, Dr. Ayol Samuels and Prof. Zvi Zemshlany, for reading and critically reviewing many of the chapters of this volume. We also thank Dr. Tzahi Ben Zion, Prof. Jonathan Benjamin, Prof. Yuval Bloch, Prof. Michael Davidson, Dr. Tom Galilee, Prof. Leon Grunhaus, Dr. Alex Kaptsan, Prof. Bernard Lerer, Dr. Arturo Lerner, Prof. Shaul Lev-Ran, Dr. Akiva Lichtenberg, Dr. Hagai Maoz, Dr. Yaakov Ofir, Dr. Yamima Osher, Dr. Nadav Reiss, Dr. Alon Reshef, Dr. Yaniv Saisky, Dr. Vadim Savlev, and Prof. Gil Zalsman, for reading chapters in their area of expertise. Remaining errors or incorrect opinions are solely the responsibility of the authors. The untiring technical, graphical and references support of Yehudit Curiel during the year of writing this book is gratefully acknowledged.

Pesach Lichtenberg wishes to express his love for his wife and dearest colleague, Bitya, in appreciation of her unwavering love and support, in this initiative and in so much more. אַהֲבָתָהּ תִּשְׁגֶּה תָמִיד - "You will always dote in her love" (Proverbs 5, tr. Robert Alter). On this occasion, he also fondly remembers his late mother of blessed memory, whom he recalls saying too often, "come on, you don't need those pills!"

Contents

Chapter 1
Introduction

This book is not a review of the literature by professors of psychiatry who do mostly data analysis. Both of us are clinical psychiatrists who have seen many patients in a wide variety of settings, and have come to appreciate controlled studies as only part of psychiatric knowledge. The book covers many aspects of psychopharmacology, but is not a textbook of psychiatry and we assume familiarity with patients, with current clinical diagnosis, with psychotherapy in its various forms, and with basic neuroscience. We write to an audience who has begun to feel the limitations of psychopharmacological claims and DSM-5 diagnoses. We hope to speak to such a reader and give him a perspective from our experience.

As senior author (RHB) I would like to explain to you where I am coming from and what might make this book relevant for the changing times of psychopharmacology. I have a fifty-year career in laboratory and clinical psychopharmacology, have been President of the International College of Neuropsychopharmacology (CINP) 2008–2010, President of the International Neuropsychiatry Association (2013–2015), Vice-President of the International Society for Bipolar Disorders (2012–2014) and President of the Israel Psychiatric Association (2015–2018). I have fully participated in many of the developments of these 50 years in the field, but feel that it is time to make clear our failures, our limitations, and our unfulfilled promises. I have struggled to avoid making this book one of the many current antipsychiatry tirades. Instead, we hope it will be a readable concise textbook incorporating up-to-date self-criticism and the scientific modesty necessary to allow the clinician and field to go forward without denying the many positive contributions we have made.

1. Genes: I started my career in the early 1970s when the heritability of many psychiatric illnesses became a proven fact. With the vision of molecular genetics on the way, we were sure that we would soon have genes [1] for psychiatric disorders and with those genes, rational targets for pharmacotherapy. None of that happened despite a huge number of high profile papers and many reviews,

© The Author(s), under exclusive license to Springer Nature Switzerland AG 2023
R. H. Belmaker, P. Lichtenberg, *Psychopharmacology Reconsidered*,
https://doi.org/10.1007/978-3-031-40371-2_1

academically thorough and convincing, that summarized the "rapid progress" and "soon to be revealed discoveries". We found no major genes even for the strongest candidate such as Tourette's. Instead, we found hundreds of common variants of small effects that constitute the heritability of mental disorders. These are no different from the many genes of small effect size that affect most traits such as height, cholesterol level, activity level, sleep depth and other normal physiological and behavioral functions. In no psychiatric illness do we have a clear cut-off for diagnosis based on a gene [2].

2. Lithium: Lithium was my first love and I was a member of one of the first groups in the USA to give this compound with an IND (investigational new drug) license at Duke Hospital in 1970 when I was a still a medical student. Sam Gershon called lithium the "first salvo in the psychopharmacological revolution". It seemed that a compound so simple with no metabolism, that exits the body the same ways it enters and that is mined in the earth rather than synthesized, must be capable of giving us a peek into the window of the mind. Lithium seemed so similar to sodium and potassium, the important ions in neurotransmission. I have worked in lithium research and read dozens of high profile papers that purported to have discovered the mechanism of lithium action [3]. All have proved to be epiphenomena, and none has led to a rational lithium-like alternative drug [4, 5]. Lithium's efficacy in many cases of bipolar disorder is undeniable. However, its specificity as a "pharmacologic bullet" for bipolar disorder is clearly untrue today. Lithium works in many but not all cases of bipolar disorder, no matter how clearly you define bipolar disorder [6]. No genetic variant has been discovered that reliably predicts lithium response. Lithium helps to augment antidepressant treatment in cases resistant to usual monoamine reuptake treatment. Lithium helps in preventing some cases of recurrent unipolar depression. Lithium helps in many cases of schizoaffective disorder. In short, lithium is not specific. Moreover, carbamazepine, valproate and perhaps some other anticonvulsants also work in bipolar disorder with overlapping indications. No common mechanism of action has been discovered. Most confusing for my own research hopes, the second-generation antipsychotics have been found to be wonderful mood stabilizers in bipolar disorder. They work for many patients and don't work for other patients. They work for some patients who are lithium non-responders and can also replace lithium in some patients who develop lithium side effects. And vice versa. Meta analysis upon meta-analysis has shown this completely different class of compounds, dopamine receptor blockers with effects on serotonin receptors in many cases, to be equally effective to lithium in bipolar disorder however one defines bipolar. This does not deny that there may be some patients who need only lithium; but there are some patients who need only olanzapine and some patients who need only valproate [7]. The recent network meta-analytic method that is the new standard shows no difference or specificity across the board between lithium, antiepileptic mood stabilizers, and antipsychotic mood stabilizers. The fact that so many meta-analyses found effectiveness for lithium and that so many academics were involved in determining its mechanism of action gave it a certain "mystique". Everyone believed that lithium represented

the ultimate proof that a holy grail exists in psychopharmacology. This belief can no longer be promoted and perhaps was harmful in the long run.

3. Antipsychotics: Early on in my career I was exposed to Solomon Snyder's elegant dopamine hypothesis of schizophrenia where every compound that benefits schizophrenia could be shown to have an effect in the patient at an average clinical dose in direct correlation with its affinity to block dopamine receptors in vitro in homogenates of rat brain using totally neurochemical analysis. This electrifying discovery supported the belief that schizophrenia is a definable illness and that antipsychotics are a definable treatment indicated whenever this illness exists. The finding that antipsychotic drugs work by blocking dopamine receptors still has truth in it. However, the effect of neuroleptics is almost entirely dependent on the effect of the dopamine blockers on positive symptoms of psychosis. Also, many patients with a diagnosis of schizophrenia recover after a few years whatever treatment they receive. Dopamine receptor blockers are not specific treatments for schizophrenia but only for psychosis. Neuroleptics are also effective in mania even without psychosis and often effective in psychotic depression. Atypical antipsychotics are good agents in severe anxiety disorder and non-psychotic agitation. The idea that dopamine receptor blockers are a specific treatment of schizophrenia may be as misleading as the idea that lithium is a specific treatment for bipolar disorder. Dopamine receptor blockers help psychosis and other conditions for reasons which we do not fully understand and which need to be carefully adjusted to the case of each patient. These drugs are not magic bullets as we would think of penicillin as a compound that only affects the bacterial cell wall and not the mammalian cell. Dopamine receptor blockers affect normal thought processes including perhaps creativity and motivation of the patient and are generally hated by our patients. Unrelated to their physical side effects, this is a price to pay and must be weighed by each individual patient with his doctor.

4. Antidepressants: The development of the SSRIs was based on research that imipramine affected both noradrenaline and serotonin reuptake and that specific serotonin reuptake inhibitors and specific noradrenalin reuptake inhibitors might give us more specific and more effective treatment. Many early convincing clinical trials showed that SSRIs were more effective than placebo in major depression. However, the definition of major depression has expanded tremendously, and has become less clearly distinguished from psychosocial stress [8]. Placebo-drug differences in research studies have narrowed to where they are almost non-existent. We all know patients who have been greatly helped by SSRIs, but these are more often classic panic disorder patients with secondary depression rather than depression sui generis. The idea that depression itself is a chemical imbalance in the brain has not been confirmed. The huge increase in the rates of use of SSRIs in the population suggests that much of its use is for life problems that do not have pharmacological solutions. The SSRIs, like the dopamine receptor blockers and lithium, may have effects on personality and social relationships of the patient. They are not magic bullets. They affect neurochemical systems that are basic to human social behavior. They should be used with care and not

with excessive enthusiasm. I have become increasingly uncomfortable with the cultural, professional and scientific promotion of SSRI as cure-alls for a chemical imbalance in the brain that we have not found.

5. Biological markers: In 1974 I participated in a group reporting that platelet monoamine oxidase (MAO) is low in patients diagnosed with schizophrenia [9]. This led to several further studies in my own lab, then in Jerusalem, and literally hundreds of attempts to replicate the finding around the world, some positive and others negative. Patients with schizophrenia eat differently, exercise differently, and feel different amounts of stress than controls. It gradually became clear that these artifacts, not schizophrenia itself, cause lowered platelet MAO. I summarized these sobering afterthoughts in an essay entitled "Lessons of Platelet Monoamine Oxidase" [10]. Biological markers may exist, but all efforts to date have not found a reliable biological way to diagnose or predict psychopharmacological treatment response.

6. Clinical trials: Clinical controlled trials, the systematic follow up of a patient with an illness after double-blind allocation to test drug or placebo, entered medicine after World War II and swept through psychiatric thinking during my residency in the early 1970s. I was captivated! Quantitative measuring scales for illnesses, clear diagnostic standards, the hope of reliable therapeutic results. I participated in many controlled clinical trials, both those sponsored by international pharmaceutical companies and those initiated by my own ideas and funded by the Stanley Foundation or NARSAD (the precursor to the Brain and Behavior Institute). Few of these studies have been replicated over the years. In commercial studies I attended training sessions on use of the Hamilton Rating Scale for Depression, the PANSS rating scale for schizophrenia, and others. I could not escape the feeling that the sessions were aimed at reliability, not validity. The commercial drugs all seemed to be "me-too" with the training sessions and double-blind methodology in place to satisfy the FDA, not to really find new drugs. Because of variability in results, meta-analysis methodology was developed. But different meta-analyses also found different answers to the question "which is the best antipsychotic for schizophrenia? " or "Which is the best antidepressant for depression?". A reanalysis of my own work led me to the conclusion that individual differences in these syndromes are so great that patient history is a far better guide to psychopharmacological treatment than formal DSM-V diagnosis [7].

7. Stimulants: I am not a certified child psychiatrist but I did do a 6 month rotation in child psychiatry as a resident in psychiatry at Duke and carried two child psychotherapy cases for a year with supervision. I have seen child patients and also children of patients continually throughout my career. Throughout my 50-year career since my MD degree I have known Paul Wender and Joseph Biederman and others who developed the concept of ADHD from its roots. I have seen many cases of adult amphetamine addiction. In the study of schizophrenia, the psychosis caused by amphetamine has been called a classic model of schizophrenia that is reversible by dopamine receptor blockade. With the widening use of stimulants for ADHD, child psychiatrists and others who advocate stimulant use

usually ignore the literature on amphetamine abuse and amphetamine effects on the brain. These two literatures are so separate that I have become intellectually and clinically uncomfortable.

A large literature exists suggesting that amphetamine and other stimulant treatment helps children with hyperactivity. I remember Judy Rappaport for many years believing that this effect is "paradoxical" and only occurs in children. I remember previous to that Wender believing that it only occurs in children with MBD and soft neurological signs. We've come a long way since that time and there are some suburbs in Western countries where a third of the children in the schools are taking Ritalin or a Ritalin-like substance [11]. This does not make sense for a specific diagnosis. I have seen with the antipsychotics and antidepressants how a literature can develop to be very convincing without examining its own premises. I wonder if the ADHD patients that appear in studies such as the MTA and others are the same patients that are so generously and thoughtlessly given amphetamines in large numbers in our schools? I am even more concerned by the large numbers of adult patients who come to me wanting Ritalin or other stimulants. Most of them seem to me to be suffering from anxiety, a sense that they are not intellectually as successful as they would want to be (that includes all of us I guess) or just dissatisfied with themselves. They all expect amphetamine to work wonders. They demand it. I see many psychoses in high profile lawyers and celebrities given amphetamines for ADHD and who come to me for their subsequent psychosis. We need to use drugs for ADHD with the care incumbent on treatments that have become so widespread and with the caution indicated by the contradiction between dopamine's role in psychosis and its role in attention and reward.

8. Our book: About a decade ago I reconnected with a student of 40 years ago and now colleague (Pesach Lichtenberg) who had researched and published on placebo effects and brought Irving Kirsch to an academic visit to discuss the diminishing placebo-drug effects in antidepressant research trials. This began a bittersweet re-evaluation for me of the accomplishments vs disappointments of the psychopharmacology of depression. Pesach and I considered whether the hypertrophied expansion of depression diagnosis was a culprit; we published a tentative proposal to restore complexity and context to depression diagnosis. Pesach went on to found a center for non-pharmacologic treatment of psychosis in Jerusalem and received massive support in the press and public tired of mental hospital overdosing and restraints. Robert Whitaker, a well-known critic of antipsychotic medication, visited Israel for a series of lectures, which produced a good deal of controversy, challenged an overly complacent consensus about our use of medication, and led many in the field to reexamine previously-held assumptions [12]. Pesach (bravely, in my eyes) invited me to be on the board of advisors of his center. I agreed, and I owe to Pesach my exposure to the last decade of literature and ferment on whether low-dose antipsychotic treatment or short term usage of these drugs may be a better path to treatment of patients with psychosis than a DSM-based lifelong prescription to all patients diagnosed with schizophrenia at dosages adjusted to stamp out any psychotic thinking. This

book is the product of our relationship, and we have worked hard to make it a true reflection of the field's self re-examination, without throwing the baby out with the bathwater. We hope this book will be a textbook for young psychiatrists about all areas of psychopharmacology and will teach in a positive way the limited use of medication when necessary in psychiatric disorder, without unproven assumptions of diagnostic specificity or biochemical etiology.

RH Belmaker, MD.

Modiin, Israel.

December 15, 2022

Thirty-six years after starting my psychiatric residency training in Jerusalem, I (PL) am rarely junior anything, but it is an honor to be a junior author when my first serious teacher in psychiatry, Professor Haim Belmaker, initiates this important reappraisal of psychopharmacology and graciously offers me to join him on this arduous publishing adventure. Over the decades, Haim has remained for me a source of wisdom and fount of knowledge. Working on this textbook with Haim has been an unremitting pleasure. Haim definitely led the way, I was happy to be at his side (mostly virtually), contributing as well as I could. If I have had some small influence on his reconsidering the role of psychiatric medication, I am gratified.

I was drawn to psychiatry because of a fascination with the workings and vicissitudes of the mind. I began my training in the field during what was a heady time for the profession. Psychopharmacology had by then become firmly ensconced as the crucial tool of psychiatrists. Psychiatry had moved from mind to brain, and many practitioners – I was among them – proclaimed that biological psychiatry is the only psychiatry, reflecting a reductionist fervor which sought to interpret deviations of mind in terms of the activity of brain. This conceptual structure ran far ahead of the evidence, but we were confident that that would change. The heuristic value of a reductionist approach was enormous. We believed that neuroscience, aided by enhanced methods of brain imaging, would soon unlock the mysteries of the black box in our skull. We were certain that advances in genetics would decipher the source of much of the emotional suffering of our patients. We were on the cusp of finally solving the mystery of mental illness, and eliminating or at least mitigating this awful source of human misery. This was what good, well-intentioned colleagues optimistically believed, and it was then hardly dependent upon the machinations of the pharmaceutical industry. And we were confident that the next, approaching sign of progress would be a new slew of medication, due to come to market in the 1990s.

Ten years later, during a brief stint as a regulator working for the local Ministry of Health, I organized a well-attended national conference on the topic of the so-called second-generation antipsychotics. A buzz ran through the lecture hall. The assumption was that these medications are more effective as anti-psychotics than the first-generation medications which they would certainly render obsolete. The clinical relief at leaving behind the problem of extrapyramidal side effects was palpable, and the attendees were discussing the medicolegal implications of prescribing a first-generation antipsychotic, lest the patient develop tardive dyskinesia and the psychiatrist be held liable for damages. The only cautionary voice I recall was Haim Belmaker's, who observed that we do not yet know the side-effects of the new medications (Eli Lilly may actually have already had some of that information by then, but managed to avoid its seeping out into the public [13]), and he suggested that until we did, we ought not rush to unreservedly embrace the new treatment options. People were not pleased to hear this message, as if Haim was ruining the party, but he was prescient. Not too long afterwards appeared the first analyses suggesting that the touted therapeutic advantages of the new antipsychotic medications were illusory [14].

Antidepressants weren't faring much better in the research arena. Irving Kirsch's early studies [15], claiming that most of the therapeutic effect of anti-depressants could be attributed to placebo effects, were met with dismay and antagonism by many in the profession, but seemed to match my own clinical experience, and have largely been corroborated by subsequent research, including a meta-analysis for which I was a coauthor [16].

But my doubts about psychopharmacology's achievements were further aroused by considerations even more fundamental than the disappointing performance of medication in placebo-controlled randomized controlled studies (RCT). For many years I ran a locked psychiatric ward, where I could follow closely the results of the medical treatments, which were often the linchpin of treatment programs. I was not pleased. The seemingly deranged minds of many of my patients were too fascinating and complex to be reduced to the blockade of dopamine receptors of the various subtypes. The medication we offered, more often than not, seemed merely to sedate our patients (when did we stop calling these chemicals "major tranquilizers?) rather than providing them with any true self-understanding. I was thankful for their calming effects, but this hardly seemed treatment, and I found it hard to believe that we were correcting any sort of reported "chemical imbalance". I got help wrapping my mind around this problem by Joanna Moncrieff's distinction between a disease-centered model of drug action, according to which the medication actually corrects an etiological cause illness (think of an antibiotic reducing pathogenic bacterial concentrations in tissue), which has been our profession's illusory dream, as opposed to a drug-centered model, where the point is the drugs' effects to provide symptomatic relief (think of a shot of whiskey to prevent pre-lecture jitters; think of an intramuscular injection of haloperidol to calm this morning's violently psychotic admission to the ward) [17, 18].

Parallel to this reappraisal of the role of all or at least most of contemporary psychopharmacology, the profession began to reconsider the level at which we should be thinking about psychiatric disorders. The 1990s was proclaimed at its start "the decade of the brain". For psychiatry, we expected that this would mean a fairly complete mapping of subjective distress onto aspects of brain functioning, with concomitant psychopharmacological remedies. But this has not happened, and perhaps should not have been expected. While the proper way of reducing mind to brain remains an active and unresolved area of philosophical debate, I think that certain conclusions can be offered, and they are necessary for the proper execution of our psychiatric practice:

1. all subjective phenomena have a neurophysiological parallel necessary for their occurrence: i.e. mind derives from brain;
2. nevertheless, minds, not brains, have subjective experiences, such as thoughts and feelings;
3. diagnosis –the fundamental determination that a person suffers from some sort of psychiatric disorder – will first of all always be on the level of a person's behavior, physical and verbal, and not on the level of the brain [19, 20].
4. the *meaning* of a person's communications is only to be found in the realm of the mind.

The latter point is worth dwelling upon. A psychiatrist's task is complex. He is a physician and must be cognizant of metabolic processes, particularly in the brain, which can produce psychopathology; he must be sensitive to patterns of speech which can imply various impairments of thought processes; but at the same time, he dare not forget that the patient, who may be psychotic, is also trying to communicate some meaning, of which he may himself be dimly aware. This last aspect can be too often neglected in contemporary psychiatry. Wherever people meet to talk, they expect that they will try to understand their interlocuter, who will be doing the same. This is not true of someone psychotic being treated today in the mental health care system. To call someone psychotic is generally taken to mean that there is no meaning to the person's utterances, and no reason to seek one: it is all detritus of excessive dopaminergic activity, to be wiped away and discarded like a runny nose. The result is that a mental patient loses his right to be heard and understood. The resulting loneliness is a harrowing result, and the deepest form of stigma. Medication enforces this image: we are helping her by giving medication. Is there meaning to what she says? Do we need to listen to what the voices say? Or to the content of her persecutory fears? Why bother even thinking about these questions, if they have no value in deciding which medication at what dosage we should be prescribing? For after all, we are treating a brain, not a mind. Over the years, I came to understand that this perception of the ground of mental illness is the most serious, most common, and most neglected side-effect of the medications we offer.

A further possible side-effect of the medications, also not to found in the little flyers which accompany the boxes of pills and contain long lists of adverse reactions, is the passivity they may encourage. I sympathize with those patients, familiar to all clinicians, who go from doctor to doctor in search of the right medication which will cure them, undeterred by all the failures they have accumulated, holding tight to the hope that they will at last be prescribed the saving pill. (Admittedly, one might approach psychotherapy in similar fashion.) A strictly medical model entails that I bring a problem with my body to a doctor, who ministers to my needs and provides a cure; I need do little more than follow directions. The patient's relationship to the treating physician is similar to that of the client bringing his auto to be repaired. He might not know the difference between a carburetor and a radiator, or what exactly the transmission does, but he trusts that the mechanic will make the proper diagnosis for that disturbing sound and repair the engine. This model is fine for fixing a car, and probably for many medical ailments, but is woefully inadequate when dealing with a person's subjectivity, where partnership between healer and patient is crucial.

And if we are discussing side effects, one must mention the deepening of stigma which can be the unintentional result of brain-based interventions for people in emotional distress. This was certainly not what we had thought would happen: quite the contrary. Along with the fanfare greeting the aforementioned "decade of the brain" came the optimistic belief that rebranding mental illness as brain disease would surely in one fell swoop eradicate stigma: why should a brain problem be any more stigmatic than lung or pancreas disease? I can still hear some psychiatrists, usually those who were already colleagues in the 1990s, spouting this misinformation. In fact, the anticipated change in attitude never materialized. Our mistake (and I was certainly amongst those who erred then) was that we did not take into account how mental illness really is different. It affects subjectivity, the inner person, in a way that disruption of any organ system does not. Ignoring this, and relating to psychiatric distress as we would to any other illness, actually seems to have deepened the scourge of stigma, at least for serious mental illness [21]. Psychosocial explanations produce better results for our patients: they arouse more empathy and compassion, and lead to less rejection and marginalization [22].

I am hardly alone in my disillusionment with psychopharmacology and its ramifications. The medical historian Anne Harrington recently wrote a damning summary of the past half-century of psychiatry which will hardly surprise those raised on the hopes of my generation: "Today one is hard-pressed to find anyone knowledgeable who believes that the so-called biological revolution of the 1980s made good on most or even any of its therapeutic and scientific promises....It is now increasingly clear to the general public that it overreached, overpromised, overdiagnosed, overmedicated and compromised its principles" [23].

There is a role for medication, and occasionally it is crucial. But psychiatry, like the rest of medicine, is a moral enterprise, where consideration of patient welfare is the first commandment of our work, while personal and corporate profit are the golden calf. This requires us to be careful in how we dispense our prescriptions, as well as other treatments. Our judgment must be informed by a sober assessment of our clinical experience and a careful reading of the evidence. This is the approach which should underly a worthy critical psychiatry [24].

Though I was involved early in my career in psychopharmacological research ("we really have great hopes for you," gushed a drug rep at the time, which kind of spooked me out), I turned my sights to a consideration of the placebo effect. An early case report which I published about a dramatic effect of placebo treatment for a physical ailment (in which I was the patient) implanted in me a lasting interest in the nature of the placebo effect and how it works [25]. Many articles discuss the necessity of neutralizing the "noise" of the placebo effect in order to get a clearer picture of a trial medication's efficacy, but what should be of even greater interest for the clinician is to understand and optimize that effect for the patient's benefit. Unfortunately, placebos cannot be patented (though I knew a researcher who came close to succeeding), which puts it at a great disadvantage in procuring funding. Ultimately, the complex, derogatorily labeled "placebo" effect encompasses all top-down processes which can be harnessed for the healing process, and are better thought of as the gamut of psychosocial interventions.

The culmination of these thoughts was the most fascinating work in which I have had the privilege to be involved, when I set up the first Soteria home in Israel [26], which has since spawned many similar homes. Soteria is a space where people, mostly acutely psychotic, come to be helped in a supportive, dialogical, non-stigmatic environment, where they are able to think about their psychosis and arrive at a better way of living with their challenges. Our homes are not anti-psychiatric by any means, but medication is just one component of a broad biopsychosocial approach, and generally not the first line treatment [27].

If the result of this textbook will be to nudge clinical psychiatry to accept the limitations of psychopharmacological solutions for the complex problems of extreme emotional distress, and to reinvigorate the search for other means of providing succor for our patients, we will have accomplished our purpose.

Pesach Lichtenberg, MD
Jerusalem, Israel
December 15, 2022

References

1. Baron M, Risch N, Hamburger R, Mandel B, Kushner S, Newman M, et al. Genetic linkage between X-chromosome markers and bipolar affective illness. Nature. 1987;326(6110):289–92.
2. Belmaker RH. One gene per psychosis? Biol Psychiatry. 1991;29(5):415–7.
3. Ebstein R, Belmaker R, Grunhaus L, Rimon R. Lithium inhibition of adrenaline-stimulated adenylate cyclase in humans. Nature. 1976;259(5542):411–3.
4. Kofman O, Belmaker RH. Ziskind-Somerfeld research award 1993. Biochemical, behavioral, and clinical studies of the role of inositol in lithium treatment and depression. Biol Psychiatry. 1993;34(12):839–52.
5. Avissar S, Schreiber G, Danon A, Belmaker RH. Lithium inhibits adrenergic and cholinergic increases in GTP binding in rat cortex. Nature. 1988;331(6155):440–2.
6. Biederman J, Lerner Y, Belmaker RH. Combination of lithium carbonate and haloperidol in schizo-affective disorder: a controlled study. Arch Gen Psychiatry. 1979;36(3):327–33.
7. Belmaker R, Bersudsky Y, Agam G. Individual differences and evidence-based psychopharmacology. BMC Med. 2012;10:110.
8. Lichtenberg P, Belmaker RH. Subtyping major depressive disorder. Psychother Psychosom. 2010;79(3):131–5.
9. Wyatt RJ, Murphy DL, Belmaker R, Cohen S, Donnelly CH, Pollin W. Reduced monoamine oxidase activity in platelets: a possible genetic marker for vulnerability to schizophrenia. Science. 1973;179(4076):916–8.
10. Belmaker RH. The lessons of platelet monoamine oxidase. Psychol Med. 1984;14(2):249–53.
11. Rubin L, Belmaker I, Somekh E, Urkin J, Rudolf M, Honovich M, et al. Maternal and child health in Israel: building lives. Lancet. 2017;389(10088):2514–30.
12. Belmaker R. Review: anatomy of an epidemic: magic bullets, psychiatric drugs, and the astonishing rise of mental illness in America (by Robert Whitaker). International network for the history of. Neuropsychopharmacology. 2017;
13. Gottstein J. The Zyprexa papers. Anchorage, Alaska: Samizdat Health Writer's Co-Operative Inc.; 2020.

14. Geddes J, Freemantle N, Harrison P, Bebbington P. Atypical antipsychotics in the treatment of schizophrenia: systematic overview and meta-regression analysis. BMJ. 2000;321(7273):1371–6.
15. Kirsch I, Sapirstein G. Listening to Prozac but hearing placebo: a meta-analysis of antidepressant medication. In: Kirsch I, editor. How expectation shape behavior. Washington D.C.; 1999. p. 303–20.
16. Khan A, Faucett J, Lichtenberg P, Kirsch I, Brown WA. A systematic review of comparative efficacy of treatments and controls for depression. PLoS One. 2012;7(7):e41778.
17. Moncrieff J, Cohen D. Rethinking models of psychotropic drug action. Psychother Psychosom. 2005;74(3):145–53.
18. Moncrieff J. Against the stream: antidepressants are not antidepressants - an alternative approach to drug action and implications for the use of antidepressants. BJ Psych Bull. 2018;42(1):42–4.
19. Schramme T. On the autonomy of the concept of disease in psychiatry. Front Psychol. 2013;4:457.
20. Stier M. Normative preconditions for the assessment of mental disorder. Front Psychol. 2013;4:611.
21. Angermeyer MC, Holzinger A, Carta MG, Schomerus G. Biogenetic explanations and public acceptance of mental illness: systematic review of population studies. Br J Psychiatry. 2011;199(5):367–72.
22. Lebowitz MS, Ahn WK. Effects of biological explanations for mental disorders on clinicians' empathy. Proc Natl Acad Sci U S A. 2014;111(50):17786–90.
23. Harrington RA. Mind fixers: psychiatry's troubled search for the biology of mental illness. New York: W.W. Norton & Company; 2019.
24. Middleton H, Moncrieff J. Critical psychiatry: a brief overview. BJ Psych Adv. 2019;25(1):47–54.
25. Lichtenberg P. Harefuah (Hebrew). 1997;132(3):167–239.
26. Lichtenberg P. From the closed ward to Soteria: a professional and personal journey. Psychosis. 2017;9(4):369–75.
27. Friedlander A, Tzur Bitan D, Lichtenberg P. The Soteria model: implementing an alternative to acute psychiatric hospitalization in Israel. Psychosis. 2022;14(2):99–108.

Chapter 2
The Biochemical Basis for Psychopharmacology in the Brain Synapse: Are We Looking in the Right Place?

The brain weighs about 1.0–1.5 kg and is perhaps the most complex system in the universe (Fig. 2.1) [1]. There are ten to 100 billion individual neuron cells in the brain. Each has a cell body and typically one long axon that projects to a distance and makes connections with other cells (Fig. 2.2). Each cell may connect to tens of thousands of other cells in a specific way. Over one hundred years ago it was debated whether the brain cells are connected in a syncytium or whether each cell has a discrete boundary. One of the first Nobel Prize winners in medicine Ramon y Cajal proved that cells in the brain have discrete boundaries. They communicate chemically (there are some exceptions particularly in lower species). The chemical connection between cells takes place at the synapse. The message that is carried from the cell body to the end of the synapse is an electrical message but not in the sense of a simple electrical current on a wire. It is the sequential opening of sodium channels along the length of the axon which leads to the depolarization of nearby sodium channels and an inrush of positive Na+ ions that then depolarize nearby channels along the axon. This propagates the message that is electrical but not a current. It travels not at the speed of electricity but at the speed of meters per second. The exact speed depends on whether the axion is myelinated, in which case the electrical depolarization "jumps" from node to node in a more rapid manner. When the transmission arrives at a synapse, the message is required to chemically cross to the next cell. At the synapse the message is carried across by neurotransmitters. The presynaptic axon whence the signal comes releases the neurotransmitter into the synapse and the neurotransmitter diffuses across the very thin and intricately made synaptic gap to receptors on the post synaptic membrane (Fig. 2.3). These receptors specifically combine with the neurotransmitter and the combined neurotransmitter-receptor complex activates a second messenger system in the post synaptic cell. This second messenger is often a G-protein connected to adenylyl cyclase. Sometimes it is a G-protein connected to the phosphatidylinositol cycle. Sometimes it is a direct connection to a voltage gated chloride channel. There is considerable

© The Author(s), under exclusive license to Springer Nature Switzerland AG 2023
R. H. Belmaker, P. Lichtenberg, *Psychopharmacology Reconsidered*,
https://doi.org/10.1007/978-3-031-40371-2_2

Fig. 2.1 Outline sketch of
the human brain

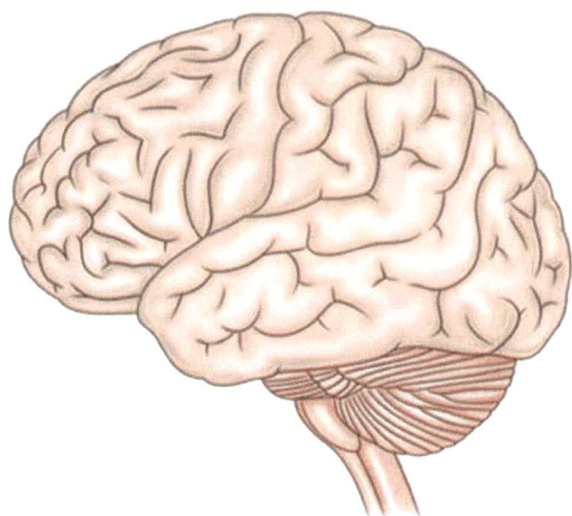

The brain ~1.0–1.5 kg

A simplified neuron

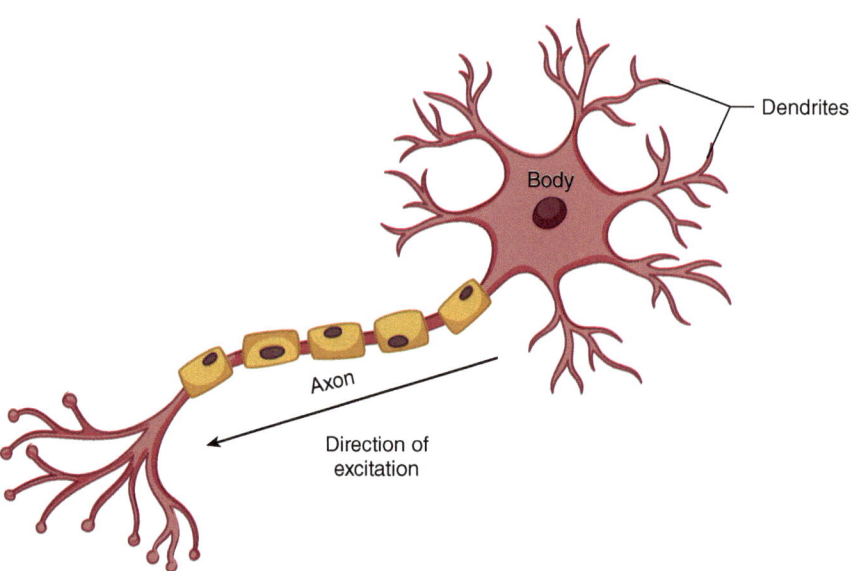

Body

Dendrites

Axon

Direction of
excitation

Fig. 2.2 Schematic of the neuron

specificity here. Each neurotransmitter connects to a specific receptor and each
receptor has its own second messenger system.

The action of the neurotransmitter at the synapse is terminated in one of several
ways. An important mechanism is reuptake by a specific reuptake pump which

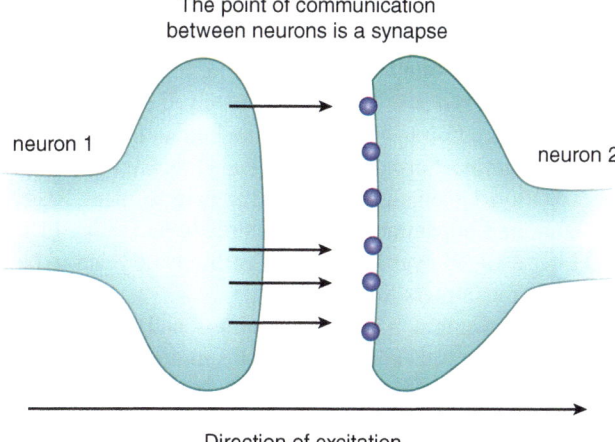

Fig. 2.3 Components of the synapse

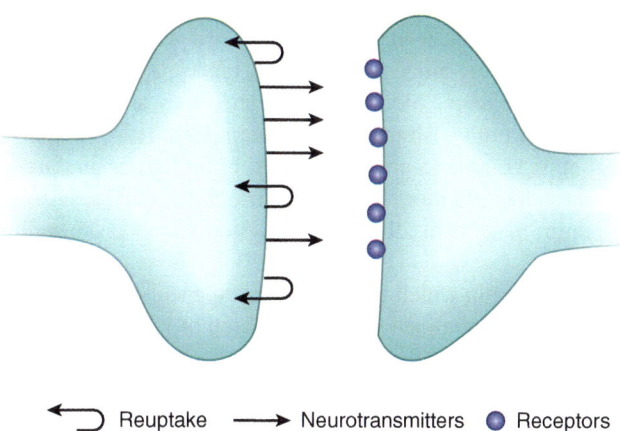

Fig. 2.4 Reuptake and receptors at the synapse

allows reuse of the neurotransmitter by the presynaptic neuron (Fig. 2.4). This pump is a target for many psychotropic drugs.

The brain system of neurotransmission across chemical synapses allows for almost infinite complexity in connection between the components of the brain. It also allows for the medical science of psychopharmacology. The chemical synapse is the site of action for most drugs that can be given to an organism in order to affect specific processes in the brain. Since there are many different neurotransmitters and many different receptors, some drugs are able to affect only some neurons in the brain without affecting others. Of course, some chemicals are less specific and affect more than one neurotransmitter and this can be the basis of side effects or multiple neurochemical effects of the same substance. However, the brain is not built as an engineer would design a complex computer. The brain evolved over hundreds of millions of years from much smaller groups of neurons that communicated in much simpler ways. The simple mollusc aplysia, for instance, which was extensively studied by psychiatrist Eric Kandel who received a Nobel Prize for his work, uses some of the same neurotransmitters that the human brain also uses: serotonin, noradrenalin and dopamine. Serotonin in aplysia is involved in eating, noradrenalin is involved in excitation and dopamine is involved in movement. Receptors for serotonin, noradrenalin, and dopamine in aplysia already had many of the amino acid sequences that later evolved over hundreds of millions of years to be the receptors used for these neurotransmitters in humans.

Other than the most common neurotransmitters and their receptors, there are many less common or less well-known neurotransmitters and receptors. Dale's rule used to state that each neuron releases only one neurotransmitter. However, there are important exceptions to Dale's rule.

The brain is surrounded by a complex blood brain barrier that is not a simple physical object that can be seen by the eye but a complex microscopic structure that ensures that most organic molecules in the blood do not simply pass into the brain. Otherwise, the food and environment of all animals would affect their behaviour chaotically. Most substances needed by the brain are transported by specific transporter systems such as for glucose. Other substances, such as lipid soluble substances, can diffuse across the blood brain barrier if they are not actively excluded. Compounds that we would wish to affect the brain for therapeutic purposes in psychiatric illness must be able to cross the blood brain barrier and are usually lipophilic. Sometimes, like lithium, they are hydrophilic but cross the blood-brain barrier using pumps made naturally for other systems such as the sodium hydrogen exchange pump.

There are other ways in which the brain is not the same as a computer designed by an engineer. The brain evolved slowly, each step of its evolution being based on the step before. So there are older sections of the brain such as the brain stem that control breathing and heart rate that were necessary even in the simplest organisms. The basal ganglia that surround the brain stem developed when more complex movements evolved, so that much of vertebrate locomotion is still controlled in the basal ganglia on an unconscious basis. The cortex that surrounds the basal ganglia in vertebrates contains most of the mechanisms for learning and the neocortex expanded greatly in mammals and the frontal cortex even moreso in primates and humans. These brain systems are not always coordinated and contradictory effects can occur in different brain systems. A psychopharmacological effect due to a drug given from outside the brain may upset brain balances and not only have unidirectional effects. There is no simple correlation between the neurotransmitter used and function of a particular brain system. Existing systems and neurotransmitters were re-used by evolution in different places and for different functions. Thus, a drug that affects the neurotransmitter noradrenaline can have effects on thinking in the neocortex while also affecting heart rate in the brain stem. Drugs that affect the neurotransmitter dopamine may affect creativity and imagination in the neocortex while also affecting posture and simple movement sequences in the basal ganglia.

This understanding is important in understanding the mechanism of side effects and their seeming inevitability. However, evolution also provided diversity in the brain and sometimes receptors that are products of protein evolution and gene duplication, became differentiated so that one neurotransmitter such as dopamine can have five different receptors in the human brain, each a product of a different gene. Attempts to find more specific drug treatment often involves the creation of blockers of specific receptors so that in the case of dopamine one might be able to block its receptors only of the D-1 type and not of the other types with hope that this might block the effects on some brain circuits while leaving unaffected its action in others.

Some old neurotransmitters such as serotonin, noradrenalin and dopamine have neuron bodies clustered deep in the brain with long axons that spread out over much of the brain. These systems have regulatory effects on much of the brain and drugs that interfere with noradrenalin, dopamine and serotonin neurotransmission at their synapses often have modulatory roles on brain and behaviour and are some of our most important psychiatric psychopharmacological drugs (Figs. 2.5, 2.6, and 2.7). Three percent of the neurons in the brain use dopamine and these will be discussed mainly in Chap. 6 with regard to antipsychotic drugs; 1% of the neurons in the brain use serotonin and 1% use noradrenalin; these will be most discussed in Chap. 5 on antidepressant drugs. About 15% the neurons in brain use gamma-aminobutyric acid (GABA) as their neurotransmitter and this will be discussed in Chap. 7 regarding anxiety. Other common neurotransmitters are acetylcholine, about 15% of brain neurons and glutamate, about 40% of brain neurons. Enkephalin or endorphin neurotransmitters are used classically by the spinothalamic pain system but are neuromodulators on many other neurons. Cannabinoid neurotransmission is used by an unclear number of receptors and is often a modulator at other synapses. However,

Fig. 2.5 The noradrenergic control system from the locus coeruleus

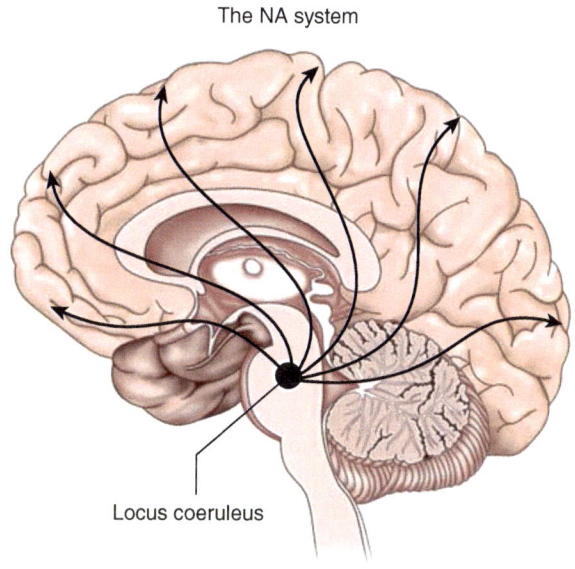

The NA system

Locus coeruleus

Fig. 2.6 The serotonergic
control system from the
midbrain raphe

The 5HT system

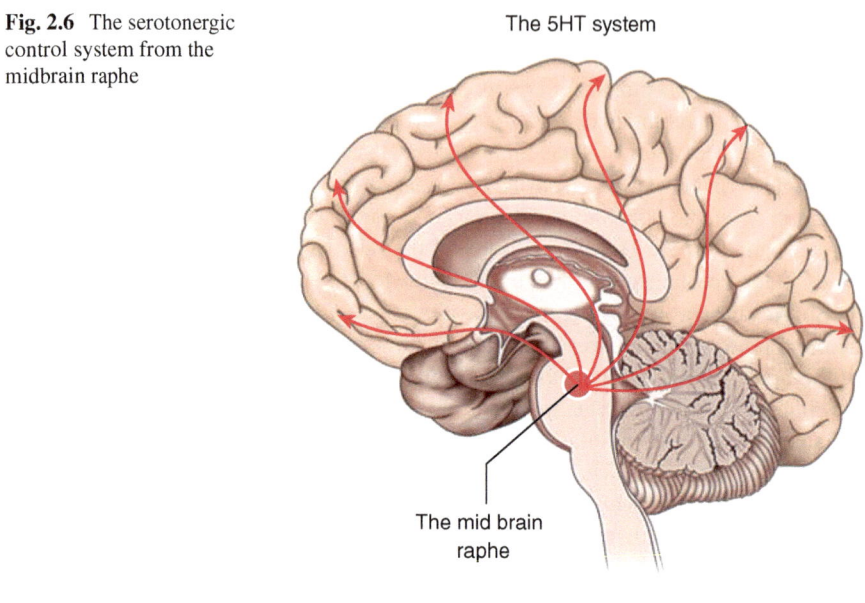

The mid brain
raphe

The Dopamine System (DA)

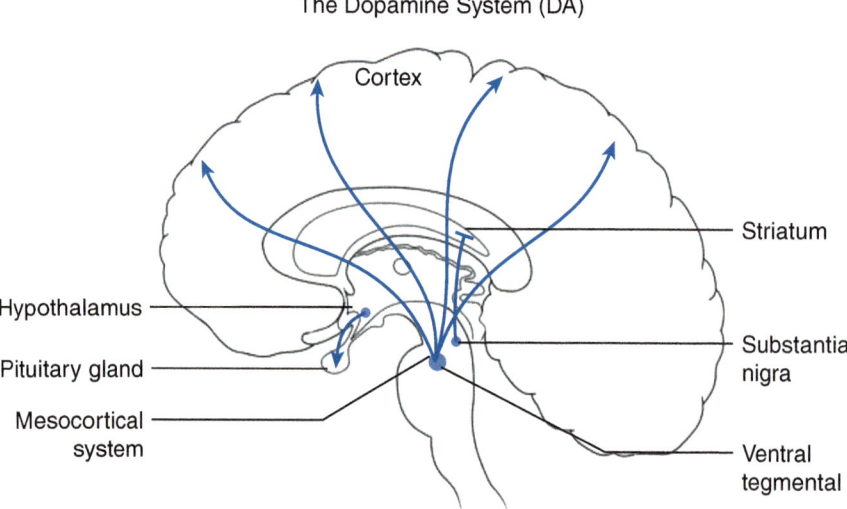

Fig. 2.7 The three major dopaminergic systems

there are probably hundreds of other less common neurotransmitters that are used
by specific neurons groups with perhaps important and as yet unknown brain
functions.

Some have criticized the concept of the biochemical basis of psychopharmacology, pointing out that these drugs can be conceptualized as medications that affect
mood, anxiety, or psychosis by affecting psychological processes such as fear

aversion, distractibility, or reward expectation. A disadvantage of the concept of the biochemical basis of psychopharmacology is its encouragement of the assumption, which has not found confirmation, that psychopathology is biochemical in origin. However, many compounds that affect attention, stimulate mood, or depress activity have not been found to be antidepressant or antipsychotic or anti-anxiety in the clinic, which weighs against a simple psychological hypothesis of the action of psychotropic drugs. On the other hand, speech and language, which might be expected to be key factors in psychopathology and cognition, do not have a specific associated biochemical neurotransmitters system. While the effects of our psychotropic substances are incontrovertibly mediated by their synaptic effects, a possible method for advancing our understanding of their therapeutic potential might be to analyze their effects upon the subjective experiences of the person receiving the medication. The jury is still out on the best model for psychopharmacology treatment and research. Jack Rhoads at Duke Medical School, a great teacher of psychiatry, used to say: "Use each model wherever it works".

Reference

1. Iversen L, Iversen S, Bloom F, Roth R. Introduction to neuropsychopharmacology. Oxford University Press; 2008.

Chapter 3
Are DSM or ICD Diagnoses the Basis for Scientific Use of Psychopharmacologic Drugs?

The American Psychiatric Association's Diagnostic and Statistical Manuel (DSM) has emerged and developed in recent years into a world-wide diagnostic classification of mental disorder. It has converged with the International Classification of Disease (ICD) section on psychiatric disorder to achieve a common language throughout the world for psychiatric discourse. The laudable aim was to facilitate psychiatric research by having agreed upon disease entities, to simplify treatment by allowing clear algorithms for proven treatments for definable diseases, and to allow for the integration of psychiatry into general medicine by using a symptom-based classification which would not require extensive psychological past history exploration of the inner life of each particular person. The major change in the direction of the new DSM approach began with DSM 3 led by Bob Spitzer. It was greatly influenced by papers published by the Washington University group by John Feighner and Samuel Guze on Research Diagnostic Criteria. The DSM-4 and DSM-5 added diagnoses and in most instances broadened them considerably [1].

It is important to note that the DSM is not a closed system such as a puzzle which is initially cut into numerous small pieces but which the puzzle solver knows when he starts working will have a definite solution where each puzzle piece will find its place. Most if not all studies of real life in patient units or patients presenting at an emergency room or clinic show that about 50% of patients do not meet any DSM criteria; that is they fall in the cracks in between the chairs [2].

The DSM movement was highly heuristic and has generated thousands of academic papers just on the issue of correct diagnosis alone. DSM diagnoses are field tested so that even before the manual is published they can be shown to have inter-rater reliability. This means that raters who are trained on video of a patient discussing his symptoms can reach very reasonable levels of agreement, perhaps 80% in common parlance or a kappa of 0.8 in diagnostic reliability jargon before a new version of DSM is approved for public use. This sounds very satisfying. However, inter-rater reliability does not guarantee validity [3]. One can define a word in the English language in a precise way while saying very little about the reality of the

© The Author(s), under exclusive license to Springer Nature Switzerland AG 2023
R. H. Belmaker, P. Lichtenberg, *Psychopharmacology Reconsidered*, https://doi.org/10.1007/978-3-031-40371-2_3

object, person, or characteristic that the word describes. The creators of the most recent version of DSM expressly wrote that they expect that each diagnosis in DSM-5 would soon be replaced by a biological diagnostic mechanism, either a DNA mutation, a physiological measurement, an electroencephalographic pattern, a PET receptor binding abnormality, or a hormonal disturbance. None of this has happened. The DSM-5 is based on symptoms just like the previous DSM and the trajectory of progress in neuropsychopharmacology does not suggest that psychiatric symptoms will be replaced by biological diagnoses any time soon [4]. Those biological findings that are so frequently reported in small studies using DSM-defined groups of "major depression", "schizophrenia", "anxiety", "OCD" etc. find significant differences in an average characteristic of genome pattern, PET scan pattern of a particular brain region, or biological metabolism. However, the significant difference is in the mean, usually with a wide variation around the mean. Most patients in each diagnostic group are far away from the mean despite the use of the strictest DSM criteria. Thus the diagnostic value of all the recent biological findings is lower than necessary for adoption as a clinical test [5]. The usefulness of a new finding or a clinical test is its specificity and its sensitivity. No biological test in psychiatry is anywhere near minimal standards of sensitivity and specificity. There were some classic flops that were widely used for a while such as the dexamethasone suppression test for depression but were then completely abandoned even by its esteemed discoverer Barney Caroll. There is no current biological test for any psychiatric disorder. Moreover, specificity and sensitivity are measured by comparing diagnosed members of a DSM disease group to controls. They do not measure the specificity and sensitivity when the disease group is compared to cases in related disease groups, a situation which is of most interest obviously to the clinician. Thus, for example, the dexamethasone suppression test might moderately distinguish melancholic depressives from normal controls but milder depressives, borderline personality, schizo-affective depressives, and persons undergoing psychosocial stress give results that are somewhat intermediate. Therefore, the test does not diagnose in a specific way the group in need of a diagnosis from the very groups from which it needs to have a differential diagnosis [6].

The DSM represented an intellectual movement in psychiatry whose roots were partially founded in the great disappointment with the psychoanalytic psychiatry that dominated post World War II American psychiatry. Psychiatrists became convinced among themselves that extensive exploration of an individual's life-long personal history and inner fantasy life rarely led to helpful results for current distress in most patients. This within-profession intellectual paradigm shift did not take place in a vacuum. American medicine in general in the 1970s and psychiatry in particular was accused of being elitist and available only for that sector of the population with financial means to afford it. It was widely stated that 98% of the psychiatrists treated only 2% of the patients. While Manhattan had a psychoanalyst on every block, states of the American Midwest had a handful of psychiatrists practicing at large mental hospitals with thousands of severe mentally ill patients. The political movement to create universal medical care coverage in the United States, or at the minimum a Medicaid program to guarantee medical insurance for the most financially distressed groups, led by necessity to treatment based on quicker algorithms and involving less time-intensive psychiatric care. The huge development of clinical psychology and social work after World War II provided a reservoir of trained manpower who could be most effectively used to supplement psychiatrists rather than compete with them and replace them with less expensive psychoanalytic care. The DSM movement served the concept that psychiatrists prescribed biological treatments based on DSM diagnoses in a medical model and could leave psychological and social treatment which were more time consuming and less remunerating to psychologists and social workers [7].

Another context of the development of DSM-3, DSM-4 and DSM-5 was the undeniable explosion of knowledge in neurosciences and genetics in the last 30 years. The mapping of the human genome, the development of accurate radioligand PET scans and MRI scans of the brain, the ability to do accurate animal behavioral analyses of rodent species along with molecular knockouts of specific genes, the development of mass spectrometry to accurately study monoamine metabolism in urine, blood and CSF- all these created an atmosphere that psychiatry's job was to make accurate diagnoses and that neurosciences would soon find the biological bases that would then seamlessly replace the symptomatic words to create DSM-5. This was promised but it never happened [8]. Neuroscience is progressing rapidly despite the serious hiccups of the reproducibility crisis. Not every paper in neuroscience that makes it into the popular press replicates. Most papers that apply neuroscience methods to groups of DSM-5 diagnostic psychiatric patients do not replicate and are forgotten within a few years. Our rule is not to believe a study until it has been replicated at least five times. But the problem is not only finding specific replicable biological findings. We do not have a reliable diagnosis for the half of patients whose syndrome does not fall into a specific DMS-5 diagnosis. Many patients fall between schizophrenia and bipolar disorder and are called schizo-affective. But many patients are not classically schizo-affective but schizo-affective and closer to schizophrenia or schizo-affective and closer to bipolar. A spectrum continuum diagnostic framework has been proposed but rejected for many technical reasons. Bipolar patients and unipolar patients cannot always be distinguished and bipolar II was created to encompass this disorder. However bipolar II patients have family histories much more like unipolar patients than bipolar I patients and treatment responses much more like unipolar patients. Personality disorders are co-morbid with most cases of bipolar disorder, most cases of major depression and most cases of OCD. Rates of both OCD and anxiety disorder are highly elevated in bipolar disorder patients even when euthymic. The problem of co-morbidity seems to have knocked in the final nail in the coffin of DSM-5 because a list of five different psychiatric diagnosis on a single patient makes DSM-5 usefulness in choosing a medication treatment limited [4].

Another major factor in the intellectual development of the DSM-5 has been the FDA and its relation to the pharmaceutical industry. The FDA prefers to control medications scientifically by demanding that a company prove efficacy for a specific medication for a specific medical disease recognized by an accepted medical authority. Psychiatry has made efforts to fit into this FDA paradigm and the DSM movement has been part of that effort. It has been successful if measured by the large number of medications that have been approved for depression, schizophrenia, mania, bipolar disorder, OCD etc. over the last 25 years. However, the effort has not been successful if measured by the number of seriously mentally ill patients who are receiving national insurance disability payments because they cannot continue to work. The newer psychiatric medications have not been successful as measured by the rate of psychiatric hospitalizations over these 25 years which has not shown the expected decline. Some of the criticisms of the antipsychiatry journalist Robert Whittaker [9] ring painfully true: While the consumption of psychiatric medication has grown in North America and as the percent of the population with access to psychiatric care via Medicaid and later Obamacare has increased tremendously, the rate of persons receiving disability assistance from Social Security, as well as the number of psychiatric hospitalizations, have continued to climb.

It is hard to find a replacement for DSM-5. Thomas Insel of the NIMH made a great effort to suggest that psychiatric research concentrate on symptom clusters that might correlate with circuit patterns in the brain rather than diagnoses. He received great blowback. Part of this is because his system might not fit into the insurance schemes available to reimburse clinical psychiatric care. But it is also true that circuit pathology, while an important research direction, does not lead to an easy "druggable" solution even if specific circuits are found to be responsible for specific psychiatric symptoms [10]."

So what shall we do with the DSM? For psychopharmacological purposes, we must limit our dependence upon DSM diagnoses. Though they may be useful for communication amongst colleagues, they are not a guide to treatment. For example, if someone is psychotic and agitated, he might be helped by a dopamine receptor blocker, regardless of whether the DSM diagnosis is schizophrenia, schizoaffective, bipolar disorder, severe personality disorder, or amphetamine-intoxication.

How then do we determine the appropriate application of psychopharmacology? We might be better guided by thinking symptomatically or syndromally rather than diagnostically. In the aforementioned example, I am treating psychosis, a validated syndromal concept, regardless of diagnosis. In another example, we might prescribe mirtazapine to help someone, not because she fulfills DSM criteria for depression, but because insomnia is a prominent cause of her misery. This approach is also behind our favorable assessment, under certain conditions, of the use of benzodiazepines (see Chap. 7). Since so much of psychopharmacology is ultimately symptomatic treatment, this way of thinking about medication might be more helpful than the DSM diagnosis for arriving at therapeutic decisions, and can allow us to prescribe according to a patient-centered empirical psychopharmacology [11].

References

1. Kupfer DJ, Kuhl EA, Regier DA. DSM-5—the future arrived. JAMA. 2013;309(16):1691–2.

2. Tondo L, Vázquez GH, Baldessarini RJ. Melancholia as a DSM-5 specifier or a separate category? J Affect Disord. 2021;282:39–40.
3. Hester L. The DSM-5's supporting works: an underused resource. Lancet Psychiatry. 2021;8(7):e17.
4. Moncrieff J. Against the stream: antidepressants are not antidepressants - an alternative approach to drug action and implications for the use of antidepressants. B J Psych Bull. 2018;42(1):42–4.
5. Ghaemi SN. DSM-5 and the miracle that never happens. Acta Psychiatr Scand. 2014;129(6):410–2.
6. Nemeroff CB, Weinberger D, Rutter M, MacMillan HL, Bryant RA, Wessely S, et al. DSM-5: a collection of psychiatrist views on the changes, controversies, and future directions. BMC Med. 2013;11:202.
7. Nelson RH. Autism advocacy before and after DSM-5. Am J Bioeth. 2020;20(4):48–50.
8. Pearson GS. DSM-5: what will it change? Perspect Psychiatr Care. 2013;49(4):219–20.
9. Whitaker R. Mad in America: bad science, bad medicine, and the enduring mistreatment of the mentally ill. Basic Books; 2019.
10. Wakefield JC. Diagnostic issues and controversies in DSM-5: return of the false positives problem. Annu Rev Clin Psychol. 2016;12:105–32.
11. Frances A. Whither DSM-V? Br J Psychiatry. 2009;195(5):391–2.

Chapter 4
The Clinical Control Trial in Psychiatry: What It Is, What It Has Been, and Does It Have a Future?

The clinical controlled trial in medicine began after WWII in an effort to put medicine on a sound scientific basis. Sometimes viewers outside of medicine understand the phrase "scientific medicine" as meaning a medicine based in biological and physiological understanding of the body where the treatments flow from understanding the mechanism of disease. This latter perception is rarely the case. Usually, treatments discovered by chance or based on only very partial knowledge of the physiology of disease are brought to a clinical trial where patients are randomly assigned to receive the new proposed treatment vs. an identical placebo. The concept of the clinical trial grew out of a historical realization that many prior treatments in medicine, even those that had been widely accepted such as blood letting or mountain air, were actually worthless if not harmful. Evidence based medicine was conceived as medicine based in controlled clinical trials and not necessarily medicine based on full physiological and biochemical understanding of the disease process, which realistically is a program only for the future. Evidence based medicine was contrasted with authority based medicine where people learned the opinions of their teachers based on their teachers' clinical experience. Unfortunately, teachers in different places often had opposing opinions and medicine based only on clinical experience was highly subject to fads and to the logical fallacies of inducing general rules from individual cases.

Controlled clinical trials in psychiatry began shortly after the discovery of the antipsychotics in the 1950s. To conduct a controlled clinical trial it was necessary to devise quantitative measures of psychiatric illness. The first was the Brief Psychiatric Rating Scale which will be further described in Chap. 6. Rating scales are also used in some other branches of medicine for controlled clinical trials such as the Psoriasis Area and Severity Index rating scale for evaluating treatments of psoriasis. So, psychiatry was not unique in requiring rating scales that included multiple symptoms, subjective ratings by an experienced clinician but without laboratory or radiological measurement and without an uncontestable method to prioritize some symptoms over others. Nevertheless, the early controlled clinical trial in psychiatry were

R. H. Belmaker, P. Lichtenberg, *Psychopharmacology Reconsidered*, https://doi.org/10.1007/978-3-031-40371-2_4

highly successful in proving the effectiveness of chlorpromazine and the other early antipsychotic drugs for the treatment of schizophrenia. Rating scale measurements declined more in patients taking the drug than in patients taking an identical placebo pill, with all the patients receiving the same standard of hospital or outpatient nursing and psychotherapeutic care. The specific treatment could thus be isolated without disturbing the ongoing care in any of its other dimensions. The controlled clinical trial also allowed a description of the effects over a time course because the trial could be conducted for 1 month, or 1 year. Rarely however could trials go on beyond a year because human beings often dropped out of trials, moved to another city, became dissatisfied, or developed a physical illness that prevented continuation in the study. Dropouts of course often occurred even after 1 week or 1 month in the shorter clinical trials. They could be handled in various statistical ways that will not be discussed in this chapter but which probably do not influence the results if the dropouts are not massive. Clinical trials where half of the patients or more drop out were understood to give much less information even if statistical methods could claim to extract such information.

The clinical trial method was expanded to antidepressant compounds when they became available and were certainly able to show the effectiveness of these antidepressants in the illness called depression as it was defined in the 1960s.

What are the problems that have developed with the clinical controlled trial in psychopharmacology as a gold standard for evidence-based psychopharmacology in 2023 going forward [1]?

1. Clinical trials of active compounds vs. placebo in the various classes of drugs in psychiatry such as antipsychotics, antidepressants, antianxiety agents, stimulants, drugs for dementia and drugs for pain, usually find little difficulty in proving efficacy for an active compound. However, head-to-head comparisons of one compound with another in the same class have almost never been helpful. Huge amounts of research, money and patient time have been devoted to finding the ideal antipsychotic, the ideal antidepressant, or the ideal antianxiety medication. Almost always the head-to-head trial finds equal efficacy. Sometimes when one compound is found statistically significantly more efficacious than another, this is not replicated in a second study which might find the very opposite effect. An attempt to remedy this solution is the meta-analysis. See below.
2. Problem number two with the controlled clinical trial is that fewer and fewer volunteers are available to participate in a controlled clinical trial once the public has become aware of the fact that active effective treatments exist. What doctor would refer his patient and what family would send their loved one to a trial of placebo vs. a new antipsychotic in the knowledge that effective antipsychotic medications exist and that the individual can be certain to be treated with an effective agent rather than having a 50% chance of getting placebo? Often the pool of volunteers became restricted to those patients in the United States with no health insurance who could only get appropriate treatment by joining a clinal trial. Often international pharmaceutical companies could not perform trials effectively in the United States and moved their trials to eastern Europe or India

where they could find volunteers who had no ability otherwise to get medical care or perhaps were being treated in a medical care system where informed consent was not understood in the same way as in the United States. Sometimes the authority of the hospital director was enough to ensure that patients would volunteer. A few years ago a study from China appeared with over one thousand participants and not a single dropout over the course of a year of a pharmacological trial. This is difficult to conceive. Another solution to the problem of a decreasing pool of volunteers was the use of pharmaceutical institutes for profit who recruited volunteers for pharmaceutical company trials. These volunteers were often "professional volunteers" and had participated for pay (!!) in numerous clinical trials before. Sometimes the same patients had participated in antipsychotic trials, antidepressant trials and antianxiety trials, each time changing reported symptomatology to try to get into the trial and receive the money he wanted, perhaps to help him buy alcohol or drugs. It is no wonder that this controlled clinical trial system began to have increasingly contradictory results [2].

3. Another problem in the clinical controlled trial system was the rise of a class of academic psychopharmacology researchers who spent 90% or more of their professional life analyzing and supervising controlled clinical trials. Often they saw none of the patients in the trial themselves. They had no intuitive or direct appreciation of whether the drug worked. They personally saw few patients outside of a select group of private patients referred to their own tertiary medical center or their own private clinic. Their statistical acumen became disproportionate to their ability to sense whether specific new compounds were really helpful in a range of potential patients. Moreover, their university careers became dependent on the publication of the studies from the controlled clinical trial they designed and supervised. The pharmaceutical companies often helped in the write up and financing and later rewarded these clinicians by generous invitations to speak about the drug that they had studied. The incentive to find positive results was great. The ability to manage conflict of interest in such a setting is impossible [3].

4. The early enthusiasm over the discovery of effective antipsychotic compounds, effective antidepressant compounds, effective mood stabilizers and effective antianxiety agents led to an understandable desire to bring these benefits to humanity. A path that can be understood in retrospect was to expand the range of diagnoses where each of these drug classes is indicated. The American Psychiatric Association's diagnostic and statistical manual became a standard for the diagnostic boundaries for inclusion in drug trials around the world. The boundaries of each diagnosis became wider and wider though the 60 s, 70 s, 80 s, 90 s, 00 s and 10 s. The number of days required for low mood and other symptoms of depression declined from weeks to days. The prevalence of these disorders increased dramatically when prevalence studies were done using the wider cutoffs as DSMIII which became DSMIV which became DSMV. However, the placebo drug difference in drug trials plummeted. It required larger and larger patient groups to prove statistical difference between active drug and placebo as the definitions widened and prevalence increased. It would seem that the active drugs were useful for a subset of patients that have now become lost in widening

blurry diagnostic concepts. Part of a solution has been to restructure clinical controlled trials and to make cutoff for "clinical effectiveness" rather than mere statistically significant differences between placebo and active drug. However, these statistical fixes have not been effective [4].

5. Another problem with the controlled clinical trials has been the FDA's insistence on licensing new psychiatric drugs only after two controlled clinical trials found effectiveness vs. placebo. It was legal for a company to do five trials with three finding no effectiveness and to keep those trials locked in its secret files. Only the two trials finding effectiveness were submitted to the FDA and this was acceptable for the FDA to license a new drug. Irving Kirsh and Erick Turner [5] were pioneers in law suits and public pressure along with Marcia Angell [6] of the New England Journal of Medicine to force companies to make their full data base available to the public and to research scientists. Using full data bases, the difference between placebo and active drug in the modern volunteer based, wide diagnostic spectrum, commercially motivated controlled clinical trials of depression has declined to a tiny effect [7].

6. In order to try to combine clinical trials that found small or contradictory effects using sometimes different measures with different means and standard deviations, the meta analysis was designed as a way to summate trials. This was an extraordinarily optimistic moment in its time. A large number of trials could be found in the literature and summated to give a result that the clinician could hope to rely on. Alas, new meta analyses found contradictions with old meta analyses. The meta analyses themselves contain statistical assumptions that were not always the same and not always easy to decide on. One meta analysis found that amisulpride was the best of all of the first generation antipsychotics, another meta analysis found that perphenazine was the best. One meta analysis found that venlafaxine was the best antidepressant of all the antidepressants then available; another meta analysis covering almost all the same studies found that milnacipine was the best antidepressant. It became clear to the clinician that meta analyses done by psychiatrists that spend most if not all of their time doing statistical analysis of results of commercially funded studies of patients who they have never seen does not give us reliable data on which to build our clinical practice [8].

Most young residents in psychiatry and indeed probably most non-researchers feel that a large sample size study with hundreds of patients is much more likely to give a clinically reliable result than a small study with perhaps a dozen patients. This assumption often discourages young investigators from performing their own trials and encourages large, multi-center, commercially supported trials where each investigator enrolls only a few subjects and the results are then determined centrally by an analytical office where the blind code is opened and the statistics analyzed by people who have never seen any of the patients. However, small trials may have different functions than larger trials [9]. An innovative idea, such as a claim that chocolate is effective in depression, might be appropriate for a small double-blind trial of 10 patients comparing chocolate to imipramine and using the Hamilton Depression

Scale (HDS) in the clinic practice of an individual young psychiatrist. A large effect of the putative new antidepressant would most likely appear even in such a small trial [10]. Even non inferiority of the chocolate to imipramine or to fluoxetine would be an important statement both for the psychiatrist himself and for the professional community. However, proof that desmethylvenlafaxine is better than venlafaxine would probably take hundreds of patients if it could be proven at all, given the biological probability that a small effect size is involved. It would be of very little interest to find non inferiority in a small clinical trial with this kind of question. If a "me-too" drug is being studied such as escitalopram after the patent protection lapsed on racemic citalopram itself, biological plausibility would be that the escitalopram would have similar antidepressant effects to the racemic mixture after appropriate dosing based on molecular activity at the serotonin reuptake site. A large scale trial with hundreds of patients to prove the efficacy of escitalopram vs. placebo to gain patent protection for escitalopram with the FDA was therefore of questionable utility and ethics. If the biological plausibility is so strong for a me-too compound to be active, a non-inferiority trial of a new compound vs. the old in less than 100 patients should prove the point [11]. Advantages on side effects or in specific subgroups could then be part of post-marketing analysis (Phase IV).

Post-hoc Analyses

Most large psychiatric clinical trials use multiple measuring scales. One scale must be defined as the primary endpoint, such as a difference in the Hamilton Depression Score (HDS) or a difference in a percentage of patients reaching at least a 30% decline in HDS or the percentage of patients reaching a HDS of 15 or less. The decision on the primary endpoint requires considerable clinical judgement and experience and may vary over time and between cultures and certainly varies between the various editions of the HDS itself. In recent years clinicians have become acutely aware that antidepressant drugs for instance may be statistically effective in large trials involving hundreds of patients with only a two to three point difference in the HDS total. This difference, while it can become statistically significant with a large enough N, may not be clinically significant. As a result many researchers in the field have been looking for definitions of clinical remission, such as number of patients with HDS less than 15 or "significant improvement" such as HDS with at least 30% reduction over the course of the trial. Sometimes the total HDS is defined as the primary endpoint but the results show only a nonsignificant effect of the drug on the primary endpoint. There is a frequent tendency to then look for possible secondary endpoints such as the effect on the item "insomnia" or the item "loss of appetite" within the HDS. These secondary analyses are a statistical conundrum and also a trap. The more the researcher looks within the data for effects he did not define as the object of his trial, the more likely it is that by chance he will find an unexpected effect. A new antidepressant, for instance, might be found to have no overall effect on the HDS but to have a positive effect on insomnia or lack of appetite. These might be real effects: in the case of trazodone, a drug originally planned to be an antidepressant, it was discovered to be a useful non benzodiazepine sleep agent. However, more often these post hoc analyses are statistical artifacts that do not replicate in subsequent studies.

Acute Trials, Prevention Trials, "Survival Trials" and Crossover Trials

Most early studies of potential psychotropic substances are done for a few weeks in order to look for a positive benefit in an economically feasible method and to thereby obtain approval to market the new compound. An example can be seen in the Fig. 4.1.

A rating scale is used, the patients are randomly assigned to receive treatment or placebo and the ratings are repeated at week one, two, three and four. If the compound is effective, a statistically significant difference between the placebo and active compound groups is found, often after only 3 weeks of treatment. A more difficult kind of study that usually happens only after a compound has reached the market is the prevention trial. Such a trial would enter patients at a particular time point, for instance on hospital discharge, with a particular diagnosis. The patients would then be followed monthly and rated on a symptom scale monthly. After 1 year or more the active compound's ability to prevent relapse of symptoms could be seen as for example in the Fig. 4.2.

A third type of trial is called a Kaplan-Meier survival analysis because it was based on illnesses leading to death such as the percentage of patients surviving with aspirin or placebo when discharged after a coronary artery bypass. However, the statistics are the same in many areas of medicine; an example for psychopharmacology would be patients studied after recovery from an index episode of mania who are maintained on an active drug or placebo. They were then seen perhaps monthly and those who had had a manic episode were rated in an "all or none" fashion as if

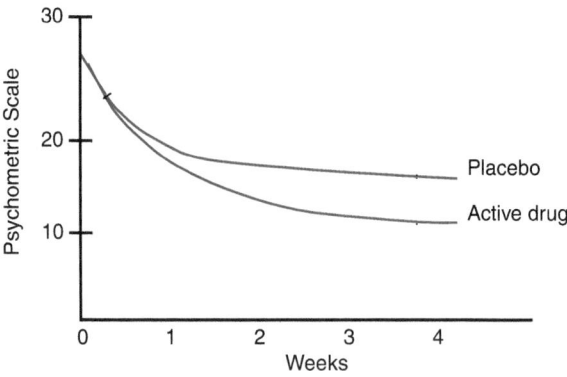

Fig. 4.1 Schematic example of the effect of placebo and active drug over 4 weeks

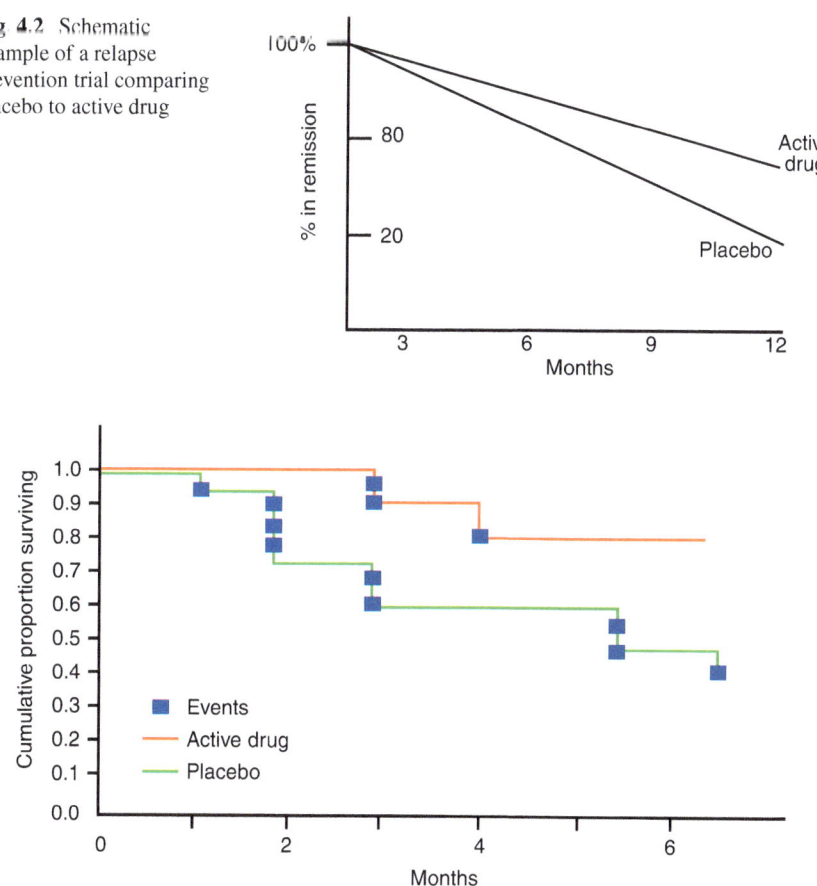

Fig. 4.2 Schematic example of a relapse prevention trial comparing placebo to active drug

Fig. 4.3 Schematic example of a Kaplan-Meier trial with a defined endpoint

they had "not survived" and each month any patient having a manic attack would thereby be excluded from further data analysis. This is illustrated in Fig. 4.3. When this kind of data is presented, often psychiatrists in the audience are confused and assume that patients who had "not survived" had developed some terrible side effect. Therefore, it is important to note here that the word "survival" is used in a statistical sense only for the kind of study where the measure is all or none.

A fourth type of trial is the "crossover trial". Crossover trials can use smaller groups of patients because each patient is his own control. The patient might get an active treatment for a week to 2 months and then a placebo control for the same amount of time. The trial must be carefully balanced so that half of the patients get the active drug first and then the placebo drug; the other half of the patients get the placebo first and then the active drug (see Fig. 4.4). The statistical analysis of crossover drugs can be more sensitive to effects in chronic illnesses such as ADHD, chronic depression or chronic anxiety disorders. However, in acute depression the

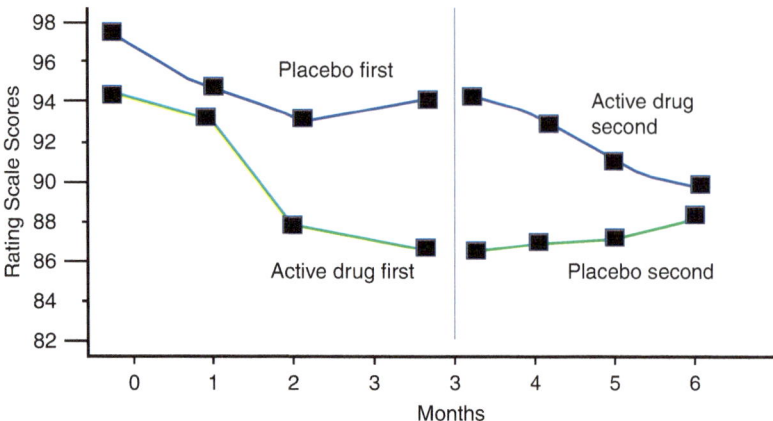

Fig. 4.4 Schematic example of a crossover trial with 3 months on each treatment

crossover trial is inappropriate and misleading because a patient who improves greatly in the active drug group during the first treatment period might not be likely to relapse during the placebo second period and thus the placebo vs active comparison would be impaired. While statisticians have reservations about crossover trials even in chronic conditions, informal crossover trials in clinical psychopharmacology where individual patients are compared over time for response to different treatments including placebo can be the most informative of any information in psychopharmacology.

The Role of the Clinical Trial in the Teaching of Psychopharmacology

It is often assumed that treatment in medicine, if it is to be scientific, is based on molecular biological findings of the mechanism of the disease which then lead to the creation of the appropriate drug treatments. As the reader of this book has probably guessed, this is rarely if ever true in psychopharmacology. It would also not be true to assume that "evidence based medicine" is merely the methodical application of the randomized clinical trial to each specific diagnosis and the creation of a list of FDA or EMA approved compounds for specific diagnoses that can then be compiled into guidelines published by the relevant professional organizations such as the World Psychiatric Association. The incorrect view is illustrated in Fig. 4.5. This teaching model is not true for reasons that we have gone over in the beginning of the chapter: (1) Almost no psychiatric illnesses have a defined biological pathophysiology, (2) Psychopharmacological substances usually found to be active vs. placebo are active in multiple clinical diagnoses but not in all patients and even then not in all patients with a specific diagnosis, (3) Clinical controlled trials often find

Fig. 4.5 An idealized but incorrect model where clinical practice is based directly on basic science mechanisms, randomized trials, and meta-analyses

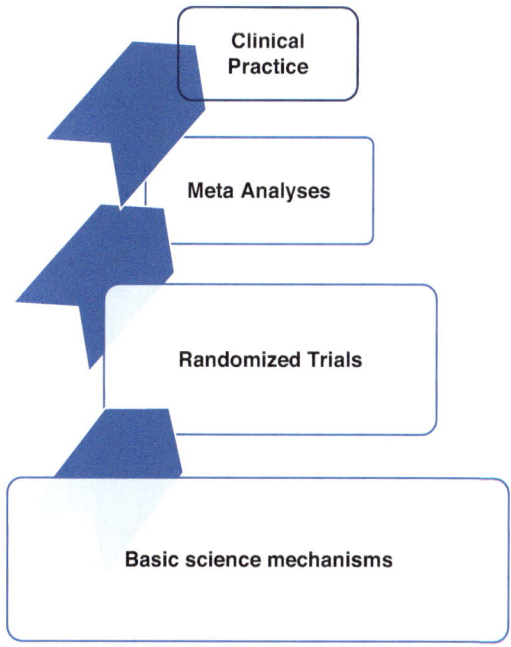

different results when performed in different settings, such as tertiary or primary care and (4) At least half of patients presenting for psychopharmacology treatment do not have a clear diagnosis or a clear evidence based treatment. There, one cannot build a structure where basic science is the foundation, topped by a layer of clinical controlled trials, topped by metal—analyses and then at the apex our clinical decision making.

We therefore present a more complex but realistic version of how psychopharmacological teaching can be based. See Fig. 4.6. As illustrated in Fig. 4.6, basic science pharmacology is informative in clinical decisions in psychopharmacology. For instance, given the accumulated number of serotonin reuptake blockers that are antidepressant a new serotonin reuptake blocker is so likely to be antidepressant that a small non inferiority clinical trial ought to be enough to make us believe that this new compound is antidepressant [12]. If it has significantly less side effects than previous ones it might be the clinician's choice even if the number of studies and statistical significance of the proofs of its efficacy were less than an older drug with the same mechanism of action. The indefinite nature of the current clinical diagnostic system means that the clinician experience base in his own practice is an important part of how he applies the literature of clinical trials to the individual patient. Thus, a pediatrician who knows that otitis media is cured by a broad spectrum antibiotic more effectively than with placebo might be more likely recommend watchful waiting if in his particular practice an epidemic of viral otitis media has been affecting a large number of his patients. He would also manage his practice for each patient based on that patient's history of previous response. His practice would be

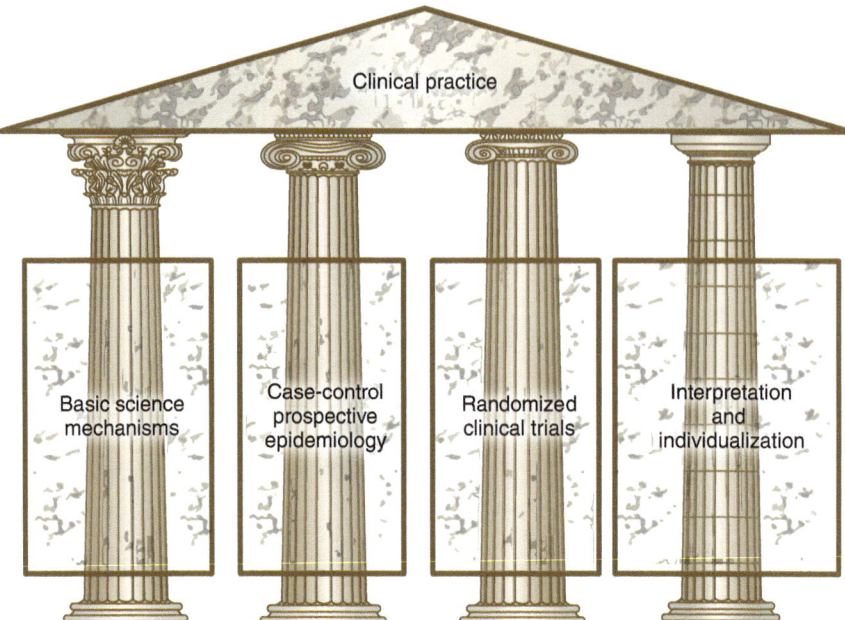

Fig. 4.6 Proposed schematic realistic model of how basic science mechanisms, epidemiology, clinical trials and meta-analyses, and individualized interpretation all contribute to clinical practice

not derived from the literature only but from experience in his particular time and place. Figure 4.6 illustrates that basic science mechanisms, epidemiology, controlled trials and also individualization by patient history and community setting all influence clinical practice decisions. No one form of input is directly on top of the other. They "hold the roof up" together is a continually changing and non-linear way.

Individual Differences

Individual differences are the basis for the philosophy of medicine as a distinct science different from biology, chemistry, physics and other sciences. Whereas in all other experimental sciences, variation around a mean is considered "noise" that obscures the truth which is the mean value, in medicine the individual patient and his individual response is the object of our inquiry and not random noise around the mean. The opposite is true, since the mean value in a medical experiment such as the efficaciousness of a particular drug treatment is really a distraction that conceals the wide variation of individual patient responsiveness. Galileo, who rolled metal balls down an incline in an attempt to discover the principle underlying an early concept of gravity, found that the balls varied one from another in the speed by which they accelerated down the incline. However, he correctly induced that the variability was due to varying wind velocities, varying surface frictions and other distractions and that the mean value asymptotically approached an underlying true physical law. In medicine it has always been clear that the results for the individual patient are not the noise but the goal of the science [13].

Figure 4.7 shows the mean differences of improvement scores between two different treatments. However, the surrounding points show the individual differences in those improvement scores which are highly overlapping. The differences in the mean may be statistically significant and with a different scientific question, this statistical difference might mean an underlying difference in a truth value. However, for the physician the individual differences implied in Fig. 4.7 means that he cannot reliably use this drug or treatment to cure every patient, or to use this finding as a diagnostic test.

Because of individual differences in patient response, clinical controlled trials have developed the concept of the "number needed to treat" NNT. This means the number of patients that would be required to take a particular treatment or pharmaceutical compound for one patient to benefit from the encounter. Statistically significant NNTs range from very strong effects of about three to weak but still significant effects of about ten. If one is talking about a relatively benign statin to lower cholesterol that might be taken by three or four patients to save one heart attack for one patient taking the compound per decade this would be good news indeed for most patients. However, if one were talking about an antidepressant that

Fig. 4.7 A difference in mean may hide large overlap of individual results

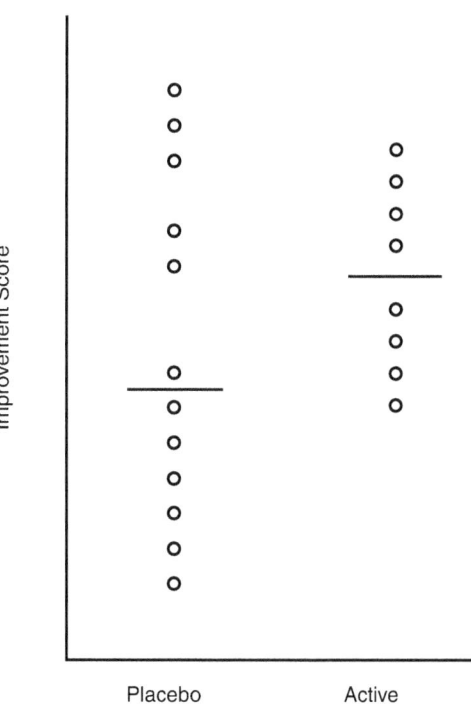

needs to be taken by five patients for one to experience a relief of acute depression whereas perhaps two will suffer sexual dysfunction, this is probably a subject that the patient and his physician should discuss openly and soberly before deciding whether this treatment is worthwhile. The answer of course will depend both on the severity of the depression, the specific characteristics of the depression and its possible response to antidepressants and the patient's degree of value of his/her sexual function.

The most clinically useful concept for individual differences is probably Bayesian Theory [14]. Bayesian Theory does not assure that there is one true answer to a question such as "does antipsychotic medication work in a diagnosis such as psychosis?". Instead, Bayesian Theory assumes that there is always a prior probability in the mind of the scientist clinician: an opinion or evaluation based on a total sum of beliefs and experiences with biology and medicine as to whether an antipsychotic might work. This is the prior probability. The Bayesian equation adjusts the prior probability as each new piece of information is added. A new controlled clinical trial showing that a new antipsychotic is effective in a statistical sense in lowering a rating scale of psychosis in a defined group of psychotic patients does not change the belief system of the medical scientist from zero non-belief to 100% belief. It incrementally in a Bayesian way alters his prior probability to a posterior probability in the direction of the new study. A subsequent similar positive study will alter the new prior probably again in the same direction. A third study, perhaps negative, will not alter his belief system from 100% back to zero but will lower the prior probability in the direction of zero to a posterior probability. The size of the change in the belief system about the probability of effect in the mind of the clinician scientist that is caused by each new study depends on the study's quality and relevance to the particular patient at hand.

Advice for the Clinical Reader of Controlled Trials

A theme of the current textbook is the understanding of psychopharmacology in a historical context. Reading medical literature exposes medical students early on to the concept of an introduction where every study is placed in the context of previous studies and existing knowledge and the reason for the conduct of the particular study is presented in each article. However, we all know that in medicine as in every area of life, the introduction and context derived from the world view and biases of the individual scientist or researcher. Some philosophers of science have provided evidence that most clinical researchers write their introduction after they have finished the study and have finished writing the conclusion. Thus, the introduction leads logically into the conclusion whether it really happened that way or not.. The authors of this volume advocate that readers of clinical trials be aware of the history of the field and the development of antipsychotic, antidepressant, antianxiety, mood stabilizers and other groups of clinical compounds. The likelihood that a new compound which is biologically a "me too" compound will be significantly better than a neurochemical close relative is small. If the new study is presented as a comparison vs. placebo, it is important that the reader be cautious in interpreting highly statistically significant differences as clinically meaningful differences.

It is also critical that the clinical reader of the research literature today not extrapolate from neurobiological significant findings in animals immediately to patients [15]. There are dozens of animal models of depression, dozens of animal models of psychosis and dozens of animal models of anxiety. Often news reports in psychiatric newsletters report on "a new treatment of depression" when they are actually reporting on a new finding in an animal model of depression. The disappointing news of the last two decades is that most of these animal findings have not turned out to be clinically effective treatments and the clinical reader should be highly skeptical about the usefulness in his practice as soon as he gets down to the section in the article and finds that it is dealing with an animal model.

Add-On Vs. Placebo Controlled Trials

As discussed above, most drug development in the 2020s is based on the already existing principles of monoamine uptake inhibition for antidepressant therapy, dopamine receptor blockade for antipsychotic therapy, benzodiazepine receptor agonism for anxiety treatment and combined dopamine receptor serotonin+5-HT2 blockade for new mood stabilizers. A new drug based on the same principles as the old drug might have a better side effect profile but is less likely to have markedly increased efficacy. To show a better side effect profile, it could be compared against an existing drug. However, the pharmaceutical companies are in a bind not of their own making. The FDA demands in the case of psychiatry that a new drug presents two placebo controlled trials for registration. The placebo controlled trials are expensive and it is difficult to recruit patients given the existence of alternative known effective treatments. Therefore, a pharmaceutical company that achieves registration of its compound with the FDA is likely to go into full scale marketing mode on the basis of its registration trials with its compound vs. placebo. This is rarely of any use to the clinician. Of much more interest to the clinician are trials where the new compound is added to an existing therapy, compared to placebo added to an existing therapy and shown to be superior in clinical effectiveness or with fewer side effects [16]. It is important for clinician organizations in psychiatry to devise an effective way to change the FDA rules in this area, which bias the field toward numerous me-to expensive compounds and to change clinical practice more rapidly than the evidence really prescribes.

The Role of Guidelines

Guidelines written by national or international committees of experts have become standards of care for many clinicians in several specialties including psychiatry. It is important for clinicians to note that these guidelines are written by committees consisting of human beings, that these experts are human beings, who also often have commercial interests and ties to pharmaceutical companies which they duly disclose but which cannot help but influence them [2]. Most patients, or at least a significant minority of them, do not fit the criterion for clinical trials. Guidelines for such patients are not akin to the model of "evidence based medicine" that we discussed that the beginning of this chapter but are more akin to the old fashioned "authority based medicine" that we discussed subsequently in this chapter. Guidelines often use a rating system where they evaluate the quality of clinical evidence for a particular drug and compare the value of that clinical evidence with the value of clinical evidence for other drugs. Clinicians mut be acutely aware of the fact that this style inherently biases in favor of new drugs against old drugs. Old drugs, a very good example of which would be lithium, were marketed after studies with small numbers of patients and using rating systems and blinding conditions that would not be accepted by the FDA today. This does not in itself make it a less good drug. Standards of blindness and sample size have been increasing in recent years, but this does not make recent drugs better: On the contrary, it shows the fact that controlled clinal trials are getting more difficult to perform and that larger sample sizes are necessary to achieve statistical significance at all [17]. Guidelines are almost always based on DSM-V diagnoses but the studies they included usually excluded patients who are suicidal, patients, who were hospitalized against their will, or patients with certain co-morbidities. Therefore, the guidelines often do not apply to real life patients [18].

Pathways for the Future

Placebo-controlled trials will always be necessary in some areas of medicine, such as oncology where a new treatment is almost always compared to placebo but as "add-on" to standard-of-care [19, 20]. In psychopharmacology, the internet promises to revolutionize clinical psychiatric research and clinical trials just as it has revolutionized so many other aspects of our lives and medical practice in particular. One illustrative case is the neurological illness amyotrophic lateral sclerosis (ALS). ALS is a relatively rare illness with a short life expectancy of a months to a few years and it has been very difficult to test new treatments in randomized trials in comparison with placebo. However, it has been possible to ask patients with ALS to report on internet sites regular assessments of their clinical status and what treatments they have received. Because of the large sample sizes obtainable from the internet, convincing evidence of the effectiveness or lack of effectiveness in ALS without prior randomization is obtainable. Statistical comparison of demographics and potential confounding variables can be performed when sample sizes are sufficient and give data that may be more useful and generalizable than the controlled clinical trial. Registration of patients with depression, psychosis or anxiety to central sites allowing them access to new compounds on the condition that they report their status at regular intervals along with their demographics and clinical history could allow accumulation of very large sample sizes that may prove efficacy or lack of it in new compounds more rapidly that the controlled clinical trial and much less expensively. Such an internet method would also avoid the increasing ethical dilemmas of the placebo controlled trial discussed above. The work of Joshua Angrist, for which he received the Nobel Memorial Prize in Economics in 2021 shows how large scale statistical exploitation of natural experiments can provide the information necessary to make prospective clinical trials unnecessary in many circumstances.

References

1. Lehrer J. The truth wears off. The New Yorker. 2010 December 13.
2. Similon MVM, Paasche C, Krol F, Lerer B, Goodwin GM, Berk M, et al. Expert consensus recommendations on the use of randomized clinical trials for drug approval in psychiatry-comparing trial designs. Eur Neuropsychopharmacol. 2022;60:91–9.
3. Paneth NS, Joyner MJ, Casadevall A. The fossilization of randomized clinical trials. J Clin Invest. 2022;132(4).
4. Mitra-Majumdar M, Kesselheim AS. Reporting bias in clinical trials: progress toward transparency and next steps. PLoS Med. 2022;19(1):e1003894.
5. Turner EH, Matthews AM, Linardatos E, Tell RA, Rosenthal R. Selective publication of antidepressant trials and its influence on apparent efficacy. N Engl J Med. 2008;358(3):252–60.
6. Angell M. The truth about the drug companies: how they deceive us and what to do about it. Random House Trade Paperbacks; 2005.
7. Ioannidis JP. Why most published research findings are false. PLoS Med. 2005;2(8):e124.
8. Bieganek C, Aliferis C, Ma S. Prediction of clinical trial enrollment rates. PLoS One. 2022;17(2):e0263193.
9. Grove A. Rethinking clinical trials. Science. 2011;333(6050):1679.
10. Bacchetti P, Deeks SG, McCune JM. Breaking free of sample size dogma to perform innovative translational research. Sci Transl Med. 2011;3(87):87ps24.
11. Bacchetti P. Current sample size conventions: flaws, harms, and alternatives. BMC Med. 2010;8:17.
12. Prinz F, Schlange T, Asadullah K. Believe it or not: how much can we rely on published data on potential drug targets? Nat Rev Drug Discov. 2011;10(9):712.
13. Belmaker R, Bersudsky Y, Agam G. Individual differences and evidence-based psychopharmacology. BMC Med. 2012;10:110.
14. Belmaker RH, Bersudsky Y, Lichtenberg P. Bayesian approach to bipolar guidelines. World J Biol Psychiatry. 2010;11(1):76–7.
15. Kafkafi N, Agassi J, Chesler EJ, Crabbe JC, Crusio WE, Eilam D, et al. Reproducibility and replicability of rodent phenotyping in preclinical studies. Neurosci Biobehav Rev. 2018;87:218–32.
16. Dagenais S, Russo L, Madsen A, Webster J, Becnel L. Use of real-world evidence to drive drug development strategy and inform clinical trial design. Clin Pharmacol Ther. 2022;111(1):77–89.

17. Anguita R, Charteris D. Could real-world data replace evidence from clinical trials in surgical retinal conditions? Br J Ophthalmol. 2022;106(8):1037–8.
18. Bhatt DL, Mehta C. Adaptive designs for clinical trials. N Engl J Med. 2016;375(1):65–74.
19. Ford I, Norrie J. Pragmatic trials. N Engl J Med. 2016;375(5):454–63.
20. Sherman RE, Anderson SA, Dal Pan GJ, Gray GW, Gross T, Hunter NL, et al. Real-world evidence - what is it and what can it tell us? N Engl J Med. 2016;375(23):2293–7.

Chapter 5
Antidepressant Drugs: For Anyone Sad or Only for the Melancholic Depressed?

The first tricyclic antidepressant, imipramine, was discovered serendipitously as a 3-ring carbon molecule similar to a phenothiazine and hypothesized to be another antipsychotic based on chlorpromazine. However, the molecule was found to have no sedating or antipsychotic properties and the clinal trial designs in that era were open enough and the patients well known enough to the investigators that the compound was noticed to be antidepressant. The diagnosis of depression at the time was far more limited than that today such that writers of the Klein and Davis textbook of psychopharmacology [1] wrote that "depression is a rare disease"! This contrasts with the depression as conceived in DSM-V and ICD-11 as major depressive disorder which affects 15 to 20% of the population at sometime in their lifetime. Depression at the time when imipramine was discovered was probably similar to what has been conceived by more recent authors as melancholia: onset over days or weeks of a state of severe depression of mood, depressed affect, sleeplessness, loss of appetite and weight, loss of sexual interest, crying, weight loss, constipation, inability to work and to function in family life, hand wringing and suicidal ideation without understandable environmental precipitation. This uncommon disease that seemed to be responsive to imipramine attracted great interest. How could the apparent effect of imipramine in this condition be measured? The answer was the development of a rating scale. One of the first used and still used today is the Hamilton Rating Scale for Depression (Table 5.1). This scale emphasized the psychophysiological accompaniments of depression such as sleeplessness and loss of appetite and excluded patients with depressive philosophy or pessimistic personality. It is not used for diagnosing depression and does not relate to precipitating events. Symptoms are rated from 0 (not present) to 4 (severe) or some modification of this rating and the ratings are summed to give a total Hamilton Depression Rating score.

Controlled trials enrolled patients with depression in the 1960s and consistently found that about 1/3 responded to placebo, 2/3 to active imipramine and about 1/3 did not respond at all. Figure 5.1 shows an archetypal result from such a study.

R. H. Belmaker, P. Lichtenberg, *Psychopharmacology Reconsidered*, https://doi.org/10.1007/978-3-031-40371-2_5

Table 5.1 Hamilton rating scale for depression

		Baseline	Week 2	Week 4
1.	Depressed mood			
2.	Guilt feelings			
3.	Suicide			
4.	Insomnia-early			
5.	Insomnia-middle			
6.	Insomnia-late			
7.	Work and activities			
8.	Retardation-psychomotor			
9.	Agitation			
10.	Anxiety-psychological			
11.	Anxiety-somatic			
12.	Somatic symptoms GI			
13.	Somatic symptoms-general			
14.	Loss of sexual desire			
15.	Hypochondriasis			
16.	Weight loss			

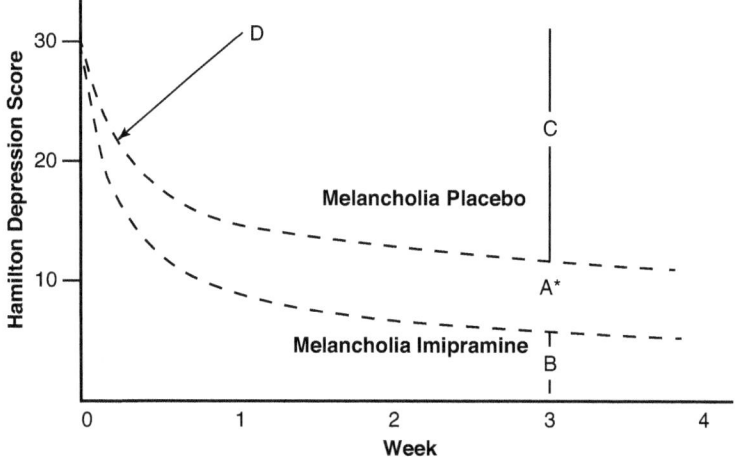

Fig. 5.1 Effect of Imipramine vs. Placebo in Melancholia

Point A: Significant differences between the placebo group and the imipramine-treated group begin at week three. This clinical point is important in guiding biochemical researchers to look for mechanisms of action that begin within the first few doses (such as monoamine reuptake blockade) versus those biochemical effects that require chronic treatment (such as changes in gene transcription).

Point B: Even after imipramine has reached a significant effect after 3 or 4 weeks, considerable symptomatology remains or a considerable number of patients remain unimproved. Many statistical analyses of antidepressant treatment have found that both are true: some symptoms remain in many successfully treated patients and some patients, perhaps are completely unresponsive in their acute depression to monoamine reuptake blocking treatment.

Point C: Many patients or many symptoms improve in the placebo group. All inpatients in pharmacological trials of this kind are receiving good nutrition, protection from conflicts with family or neighbors, treatment of concurrent medical conditions that may have been neglected, and nursing and psychological treatment. Thus, the placebo group response represents the total response to all other treatments other than the compound under study, and should not be considered a response to suggestion or expectation alone. The passage of time also heals for many and reflects the course of illness. Most depressed patients eventually improved in the pre-antidepressant era and thus for many improvement with placebo merely reflects the natural course of illness.

Point D: There is a strong tendency for both the placebo group and active treatment group to improve by week one and perhaps before, if a rating scale was done earlier, but this improvement is common to both groups and does not represent a specific effect of imipramine.

These classical results are a matter of little dispute today. However, many clinical studies have found difficulty in defining depression: Was a precipitating event such as divorce, bereavement, loss of job, medical illness, or refugee status an exclusion factor for a biological or endogenous depression? Evidence was not found and lack of evidence was taken as evidence to drop this concept of endogenous vs reactive depression and thus to greatly widen the concept of depression. Was severity of depression a necessary criterion? Again, clear evidence was not found and severity was gradually loosened as an inclusion criterion, again widening the definition and increasing the apparent prevalence of depression. These academic studies did not occur in a cultural vacuum but took place in the 1960s, 1970s, 1980s and 1990s when new imipramine-like antidepressants were introduced yearly, when the Western public became increasingly secular and eager for scientific cures for all ills, when basic neuroscience was exploding with exciting and promising new discoveries, and when pharmaceutical company profits from antidepressant prescriptions were soaring. The history may be comparable to that of opioid anti-pain medications (see later chapter in this textbook) where pharmaceutical developments, pharmaceutical advertisement, and academic research on patient needs and diagnoses interacted in not always justifiable ways.

The authors frequently meet clinicians who report that they use antidepressants freely in all patients with low mood of any kind and are very pleased with the results in their practice [2]. Unfortunately, clinical practice is not a reliable guide for understanding treatment efficacy in illnesses with a high placebo response rate and/or a high rate of remission as a natural course of illness. If most patients present with transient life stresses, these go away in any case within a few weeks. If an antidepressant is prescribed, it can take the credit. This is illustrated in Fig. 5.2.

Figure 5.3 shows Figs. 5.1 and 5.2 superimposed. Actually, melancholia while responding significantly to antidepressants has a poorer prognosis than stress

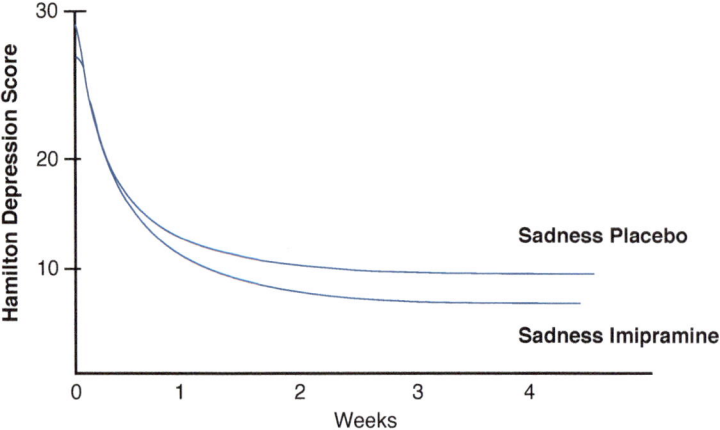

Fig. 5.2 Effect of imipramine vs. placebo in sadness, reactive depression or life stress

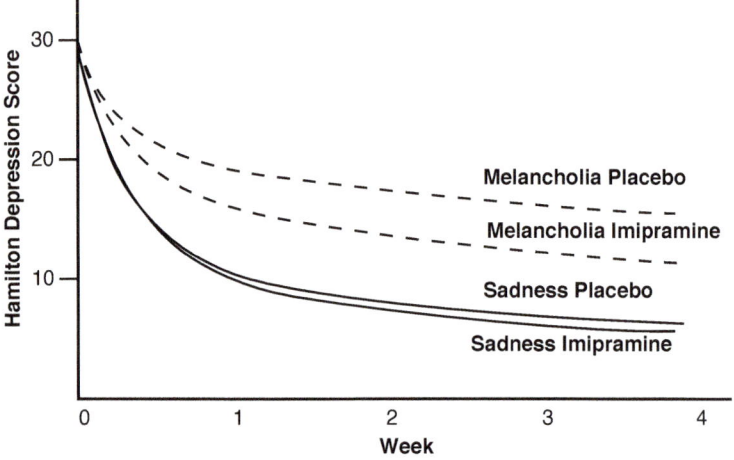

Fig. 5.3 Why clinical experience makes it seem like imipramine is at least as good for life stress as for melancholia

responses including sadness that are precipitated by life events. Thus, clinicians will form a strong impression that their "best" patients, those most responsive to antidepressants in the office, are those cases of mild depression or depression precipitated by life events. The dangers of such a conviction include the fact that antidepressants are expensive, that they have known side effects, that they may have as yet unknown side effects, and that naïve pseudoscientific beliefs undermine public trust in scientific medicine.

Maintenance Treatment with Antidepressants

Within a short time after the early acute studies of imipramine and other antidepressant drugs, the issue was raised as to whether patients who had been stabilized on these drugs need to be maintained continually on the medication or whether they had in any sense been "cured" and can go off the medication. A different design was developed to answer this question and an archetypal study is shown in Fig. 5.4. Patients were admitted to study when they had reached remission by some definition, and usually after a minimum of two previous episodes. They were then randomized as in the acute studies to imipramine tablets or placebo. Patients were usually followed not with a rating scale such as Hamilton Depression Scale but followed with a non-continuous variable such as relapse. Most studies lasted a year and some lasted 2 years. Typically, imipramine-treated patients had a relapse rate of about 20% by the end of 1 year whereas placebo treated patients had an 80% relapse rate as can be seen in Fig. 5.4 [3]. As with the acute effects of antidepressants, these effects to prevent relapse become much smaller if diagnosis is expanded to include all persistent low mood [4].

The Development of SSRIs AND SNRIs

Imipramine caused intense excitement in the psychiatric community because it clearly did not cause parkinsonian symptoms and was working on mechanisms different from those of the antipsychotic drugs from which it had been developed. It had already been noted that reserpine, an ancient Indian plant derivative used to treat high blood pressure, often caused depression in patients. Neurochemical studies found that reserpine depleted monoamine neurotransmitters in the presynaptic cells of the synapse, particularly noradrenaline and serotonin (although also dopamine). Imipramine was found to inhibit the reuptake of noradrenaline and serotonin by the presynaptic reuptake pumps and thereby to increase the amount of noradrenaline and serotonin in the synapse. The summation of these two facts, (1) the depressogenic effect of reserpine which depleted monoamines and (2) the antidepressant effect of imipramine which increased monoaminergic neurotransmission led to the

Fig. 5.4 Prophylaxis of recurrence after a depressive episode

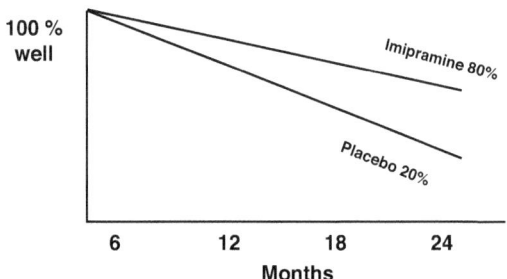

monoamine hypothesis of depression. Schildkraut formulated this hypothesis in North America with an emphasis on noradrenaline and van Praag formulated a similar hypothesis in Europe with an emphasis on serotonin. There then began a parallel and interacting history where academic psychiatric researchers designed experiments to see whether serotonin or noradrenaline was more essential in the pathophysiology of depression; and meanwhile pharmaceutical companies developed compounds that were increasingly specific for serotonergic or noradrenergic reuptake inhibition with the hope that they might be more effective in treating depression or perhaps more specific in treating some kinds or particular groups of depressed patients. The first totally specific serotonergic reuptake inhibitor to succeed in the medicinal market was Prozac (fluoxetine) by Lilly. Fluoxetine was clearly free of the anticholinergic side effects that had caused dry mouth, dry skin and GI constipation in the early depressed patients treated with imipramine. However, it is not clear if there was ever any empirical basis to Lilly's claim that its specificity to the serotonergic system led to increased efficacy. As described by David Healy in The Antidepressant Era [5], psychiatry as a profession was ready, willing and even desperate for a new answer to depression at the time that Prozac was launched. It was not difficult to convince the willing psychiatric profession and one might even say the willing public to accept a pharmaceutical claim that Prozac was far superior to the existing and by then standard antidepressant imipramine. No controlled trial has ever found fluoxetine superior to imipramine. It certainly had fewer side effects when the early antidepressant side effect rating scales were used, which measured the classical anticholinergic side effects of imipramine. However, once rating scales began to note the side effects of specific serotonergic reuptake blockade, which include nausea, diarrhea, reduced sexual desire in men and women, delayed orgasm in women and men and reduced erectile function in men, it seemed that the total amount of side effects with fluoxetine are equal to those of imipramine.

Commercial promotion of Prozac and the new me-too serotonergic reuptake inhibitors that constituted the class of SSRIs was immense and has been documented in several books about the history of psychopharmacology. The number of patients taking SSRIs in Western countries at any one time has been estimated as 10% of the adult population. Meanwhile, in a parallel fashion the diagnosis of depression from DSM-II to DSM-III to DSM-IV to DSM-V has expanded tremendously. The emphasis in an extraordinary number of publications was to identify increasing prevalence of depressive symptoms in psychiatric practice patients and in general family practice patients. The central issue of sadness, unhappiness and reactions to stressful life events soon came to be equated with the clinical diagnosis of depression. Melancholia as an imipramine responsive disorder in 1969 [1] was a very different problem than depression diagnosed with DSM V or ICD 11. As is beautifully documented in the book the Loss of Sadness by Horwitz and Wakefield [6] all emotional reactions in the human condition to life events such as unemployment, divorce, failure in university and disappointment in romance all seem to cause responses in the human psyche and physiology that could be diagnosed as major depression. Since these conditions affect the majority of human beings at some time in life, there seems to be an almost limitless potential market for SSRIs.

A scientific problem occurred, however. Studies in this greatly enlarged patient group of major depressant disorder increasingly found smaller differences between placebo and active drug in controlled studies. Companies with new SSRIs had to recruit hundreds of patients to find a statistically significant difference of 2 to 3 Hamilton Depression Rating Scale points and secure FDA approval of their drug as an antidepressant. One strategy that companies could use would be to have strict criteria to define depression in the phase III drug trials necessary for registration; but then to use much wider criteria for depression when marketing their drug as a solution for symptoms that could often be human sadness rather than depression. Kirsch [7] found that if unpublished studies were included, recent studies of placebo vs. antidepressant probably find no difference in the average patient. Moreover, because of publication bias, the published literature may overestimate the effectiveness of commercially available preparations [8].

Kirsch's claim that antidepressants are merely placebos is unlikely to be true for several reasons: First, early studies of imipramine in narrowly defined melancholia were solid evidence. Second, animal models of depression such as the Porsolt swim test find that the drugs that we call antidepressants in patients and which can be biochemically characterized as monoamine reuptake inhibitors, seem to have consistent and general effects in the Porsolt swim model and other models of depression in animals. It is unlikely that they are inert substances. Three, a more likely explanation of the decreased placebo drug differences over time is the widening of the diagnosis of depression and its use to cover such a wide gamut of human sadness that its usefulness has been diluted. Four, the evidence for monoamine-reuptake inhibiting drugs in the prevention of depression relapse is much stronger than the data on such drugs in the treatment of acute depression. Why might antidepressants be more clearly useful in prevention of relapse than in the treatment of acute depression? Studies of prevention usually include only patients who responded to the drug in their acute phase and who have had at least two episodes in the past. Thus they more closely approximate the narrowly defined melancholia for which antidepressants are often effective.

The Monoamine Theories of Depression

Figure 5.5 illustrates Shildkraut's monoamine theory of depression. This theory has been highly heuristic and generated numerous studies, from looking at metabolites of noradrenalin in urine (MHPG, 3-methy 5-hydroxyphenyl glycol) to post mortem studies of noradrenaline receptors in the autopsy brains of depressed individuals or suicide victims. As reviewed in Belmaker and Agam [9] none of these studies has consistently found a decreased level of serotonin or noradrenaline, its receptors or its reuptake sites in the brain of depressed patients.

Parallel to the development of the SSRIs, several companies successfully brought to market noradrenalin specific reuptake inhibitors such as reboxetine. The noradrenaline reuptake inhibitors are no more and no less effective than the serotonin

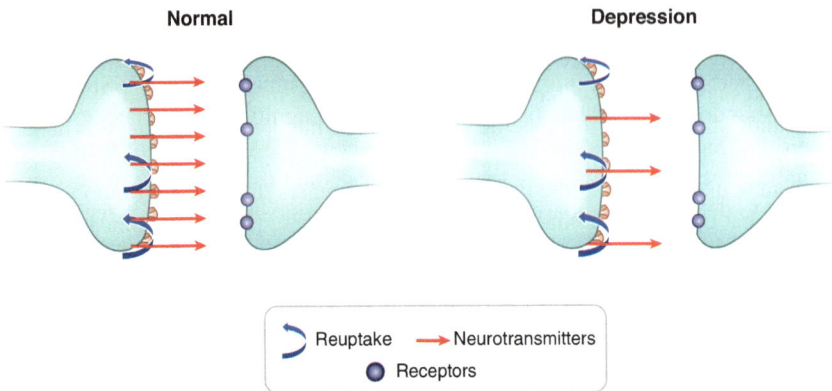

Fig. 5.5 Neurotransmitter release and reuptake at the synapse in the normal state and as hypothesized in depression

specific reuptake inhibitors in depression. They did not receive the marketing success that the SSRIs did. In about 2000 Lilly developed duloxetine, a combined noradrenalin and serotonin reuptake inhibitor, that was heavily marketed and is certainly no more and no less effective in clinical trials than SSRIs or NSRIs. Duloxetine was reinventing the wheel and returning to imipramine although it does have the advantage of not having the anticholinergic side effects of imipramine. Looking at the history of antidepressant drugs over the last 60 years from the discovery of imipramine, it seems that every year a new antidepressant was introduced and claimed to be more effective and with fewer side effects than the one of the previous year. If this were true, it must be the case that our current antidepressants are extremely successful [10]. No one claims that to be the case and all recent reviews bemoan the fact that many patients do not respond to current antidepressant therapy, that the delay in response to treatment is still 2 to 3 weeks and that side effects are still a major reason for discontinuation. Most thinkers in the field believe that it is time for a reconsideration [11]. Perhaps it would be better if the use of antidepressants is limited to those narrow group of melancholic depressives who clearly need a biological treatment and the psychosocial treatments be first line for most other patients suffering from sadness due to life events. To allow such a range of clinical practice, a splitting up of the diagnosis of major depressive disorder may be necessary as described by Lichtenberg and Belmaker [12] and the diagnosis of major depressive disorder may need to be deconstructed. More is said about this in Chap. 3.

A creative idea inspired by the ambiguity of imipramine's actions which included effects both on noradrenalin and on serotonin was that some patients might have serotonergic depressions and other patients might have noradrenergic depressions. Perhaps patients could be subtyped biochemically based on their spinal fluid 5-HTIAA (5 hydroxy-indole acetic acid) the major serotonergic metabolite that appears in spinal fluid, or urinary MHPG (3-methoxy-5-hydroxy-methylgycol) the major noradrenalin metabolite which is more easily accessed in urine. Many studies were based on this idea which seems to have failed. It is unfortunately not possible to send a urine or spinal fluid sample to a laboratory to diagnose which kind of anti-depressant a patient needs, even though the serotonergic systems and noradrenergic systems are so clearly different in the brain. For that group of patients with melancholic depression who need an antidepressant, it does not seem to matter whether they are prescribed an antidepressant to increase noradrenaline or an antidepressant to increase serotonin.

There have been further creative ideas in this field which often blossomed in academic research laboratories, were then applied by medicinal chemistry synthesis divisions in pharmaceutical companies, and were then brought to clinical trial in studies versus placebo in depression. For example, a receptor called alpha-2 adrenergic and another receptor called serotonin 1a exist on many if not most monoamine presynaptic neurons. These are self regulatory negative feedback receptors such that if too much monoamine neurotransmitter is secreted, they are stimulated and reduce secretion. Inhibiting the alpha-2 adrenergic receptor or the serotonin 1a receptor could theoretically enhance the effect of monoamine reuptake blockers to augment monoamine neurotransmission. The former idea contributed to the development of mirtazapine and the second in vortioxetine. Both received regulatory approval after being shown to be better than placebo (by a few Hamilton Depression Scale points) in controlled clinical trials. But neither has turned out to be better in head-to-head trials than older antidepressants. Their development was not the result of a profit hungry cabal but of a very creative and interdisciplinary Manhattan project, not too different from the recent effort to make COVID-19 vaccines. Marketing of these new antidepressants, however, is another matter: Their marketing has made claims unproven in firm clinical research and based only on the theoretical hopes that generated their development. Sometimes the scientists involved in development were also co-opted into the delusion, since as Nobelist Kahneman and others have shown, we all need to reduce cognitive dissonance and some do this by believing in the hypotheses they have based their research on, whether they turn out to be true or not. Clinical psychiatrists must carefully assess the clinical evidence, not the theoretical hopes. The human brain and its monoamine synapses are so complex that there is reasonable hope that future research of this kind may indeed find more effective antidepressants, and if they arrive they must find us hopeful and not cynical. Meanwhile, the psychiatrist should choose two or three antidepressants, learn their dosages and side effects, use them when indicated and go on to other treatments if they do not work in the individual patient.

Monoamine Oxidase Inhibitors

A rarely used separate class of antidepressant compounds are the monoamine oxidase (MAO) inhibitors. These were discovered serendipitously at about the same time as imipramine because iproniazid, an antituberculosis drug, seemed to improve mood in some patients being treated in tuberculosis sanatoria. The MAO inhibitors block the breakdown of the monoamines serotonin, noradrenalin and dopamine in the brain and increase the levels of those neurotransmitters in the synapse and thus their antidepressant efficacy is consistent with the monoamine hypothesis. The MAO inhibitors also inhibit the breakdown of monoamines in the liver and have the potential side effect of causing a hypertensive crisis in the event that a patient eats a food such as chicken liver, old wine, or old cheese containing tyramine, a potent hypertensive amino acid usually metabolized rapidly in the liver. Much ink has been spilled as to whether this side effect is common enough to warrant disuse of these compounds or whether a warning in the medicine instructions to patients would suffice. In practice use of these compounds, the modern versions of which are isoniazide, phenelzine and tranylcypromine, has declined so much that the average psychiatric resident and psychiatrist in the US has never used a medicine of this class. This could be unfortunate, since patients resistant to imipramine might benefit from an MAO inhibitor with its different mechanism of action. Also, consistent clinical reports suggest that MAO inhibitors are more effective than imipramine in atypical depressions and in chronic depressions [13].

Bipolar Depression

The psychiatric literature is amazingly contradictory as to whether antidepressants are useful or not in bipolar depression [14]. Most USA controlled studies find that antidepressants are of little or no therapeutic value in bipolar depressed patients and that they simultaneously carry a significant risk of switching the patient into a manic state. By contrast most European controlled studies find that antidepressants are therapeutically as useful in bipolar depression as in unipolar depression and carry little risk of switching the bipolar patient into the manic phase. How could this geographical scientific contradiction be understood? Most likely it reflects different diagnostic patterns of depression between USA and Europe. In the US diagnosis of depression has expanded so widely that many bipolar patients, who are after all also human beings with lives, employment, romances and needs to buy houses and to raise children, often have life stresses, depressions that are basically sadness or reactions to their environment and not part of their bipolar illness. These depression-like syndromes are unlikely to be responsive to antidepressants, as we have said before. In Europe by contrast depression remains more narrowly defined and bipolar depressed patients may be more likely to be having an episode of their bipolar illness. In Europe diagnosed depressive episodes in bipolar disorder seem to be more likely restricted to the melancholia that was and still is found to be responsive to antidepressant treatment.

In addition to the above theoretical discussions, in practice bipolar depressed patients must be carefully distinguished by history from unipolar patients and their treatment with antidepressants, if indicated, be carried out with mood stabilizer coverage (see chapter on mood stabilizers) and with careful frequent monitoring for early detection of manic switch.

Antidepressants in Panic Disorder

In the mid 1960s Donald Klein published a seminal paper defining an imipramine-responsive subtype of anxiety [15]. He described dramatic responses to low dose imipramine in a group of patients with sudden onset of psychophysiological anxiety symptoms including sense of heart pounding, shortness of breath, quivering of knees, dry mouth and a sense of impending doom that could occur several times per week and last for minutes or hours and had an onset in young persons with no apparent precipitating cause. It is hard for psychiatrists in 2023 to realize that panic disorder did not exist in the diagnostic nomenclature before the realization that pharmacology had carved out a treatment responsive subgroup. These patients in Klein's description responded dramatically to low doses of the then recently discovered antidepressant and enabled long lasting prevention of the acute anxiety attacks. There was no effect if taken during the anxiety attack itself. Later studies as with depression extended greatly the epidemiology of panic disorder and with a modern definition this is a very common disease rather than the rare syndrome that Klein described. The syndrome in modern diagnostic systems is not clearly differentiated from general anxiety disorders and the responsiveness to antidepressants is less rapid and less marked in modern studies than in that classic study. It could well be the case that the same historical process has occurred with anxiety that has occurred with depression; that is that widening diagnostic boundaries has led to decreasing efficacy of drugs that were effective in a well defined narrow syndrome. However, antidepressants definitely are effective in clearly defined panic disorder syndrome and as with depression there is no distinction between serotonin reuptake inhibitors vs. noradrenaline reuptake inhibitors. The effectiveness of antidepressant drugs in panic disorder suggests that depression and anxiety are closely related and often co-exist. The antidepressants should probably be called monoamine reuptake blockers rather than antidepressant drugs. It is difficult to know what fraction of patients filling prescriptions for antidepressant medication are being treated for depression and what fraction for anxiety.

See the chapter in this textbook on OCD for a surprising exception to the equivalence of SSRI and SNRI. See also the chapter on anxiety for the relationship between GABA based treatments and monoamine reuptake blockers in anxiety.

Switching Antidepressants in Treatment Resistance

The STAR-D series of clinical trials [16] was designed by the NIMH after numerous new pharmaceutical company sponsored trials had shown efficacy for a multitude of new antidepressant compounds. Since almost all of the studies proved efficacy vs. placebo, psychiatrists were at a scientific loss to decide what to do with those patients who responded inadequately or not at all to their first antidepressant. Was there a positive benefit in switching to another antidepressant? Was it preferable to switch from a noradrenergic antidepressant to a serotonergic antidepressant? Or best to begin with a dual action noradrenergic/serotonergic mixed antidepressant? STAR-D examined this question in a systematic way in a very large patient sample. Very little effectiveness was found for switching antidepressants in those patients who failed the first and for those patients who failed two trials there was almost negligible efficacy for trying the third antidepressant. This information has been difficult to digest and many clinicians continue to employ a wide variety of "switching" tactics to manage their patients who failed to respond rapidly and completely to the first antidepressant. The research community has reluctantly come to grips with these data which has been widely replicated in different settings. The psychiatric research response has been to reconceptualize depression from the self-limiting and rare disease that it was in the 1970s and from the antidepressant responsive good prognosis disease that it was in the literature in the 1990s to a new concept of depression as a highly chronic poor prognosis disease. This change may be merely a reflection of changing diagnostic patterns and should not obscure the remaining and well proven usefulness of imipramine and all of the subsequent conjoiners in clearly defined melancholia-like depression. It is also important to remember that medicine is an art and some patients with sadness unresponsive to psychotherapy and with absence of precipitating life events may benefit from a therapeutic trial of antidepressants.

Cortisol and Depression

Cortisol is a hormone secreted in all higher animals by the adrenal cortex gland in response to stress. Physical injury and infection can increase cortisol levels. Psychological stress is appreciated by the human cortex of the brain as a threat to the individual's life, integrity or social position and results in a secretion of corticotropin releasing hormone (CRH) by the hypothalamic neurons onto the pituitary gland, which in response releases adrenal corticotropin releasing hormone (ACTH) into the bloodstream which stimulates the appropriate receptors of the adrenal cortex to release cortisol. Cortisol orchestrates a complex metabolic response in the human body that facilitates successful "fight or flight" responses. Since depression and sadness are commonly associated with stressful life events and sadness is reliably related to stressful life events in research data as well, it would be reasonable to assume that cortical metabolism might be abnormal in depressed patients. Often small elevations have been reported in plasma or in saliva of depressed patients but with great overlap between the depressed and control patient groups. Moreover, increases of cortical levels in a parachutist about to jump from a plane or in psychiatric residents about to present a grand rounds seminar are higher by several fold than any increased cortisol reported in clinical depression. The cortisol system is diurnally variable with cortisol levels being higher in the morning and thus numerous studies have looked at the circadian rhythm of cortisol in depressed patient with no clear results. Some pioneering studies looked at the ability of exogenous synthetic cortisol (dexamethasone) to depress secretion of cortisol as a marker for the sensitivity of the brain receptors involved in regulating cortisol homeostasis. In about 50% of severe melancholic depressed patients, an abnormal lack of suppression of endogenous cortisol was found. This was called the dexamethasone suppression test (DST). However, similar defects in cortisol regulation were found in numerous other psychiatric disorders as well as in physical illnesses such that this test has been discarded as not useful in the diagnosis of depression.

While it would make intuitive sense for illnesses such as psychiatric disorders to involve abnormalities of cortisol the stress hormone, it is important to know that no measurement of cortisol metabolism is of use today in psychiatric diagnosis or psychopharmacology. Pharmacological blockade of cortisol receptors, which are well developed clinically, are of no use in depression treatment. Cushing's disease is not reliably associated with depression, nor is Addison's disease, although they both carry some increased risk. Intravenous cortisol may cause mania as well as depression. Large numbers of patients treated with prednisone or other cortisol like compounds for asthma or many other medical conditions are at only slightly increased risk for depression but also at increased risk of manic reactions. In post traumatic stress disorder, many studies report changes in cortisol levels but some studies report increased levels and other studies report decreased levels. This complex important physiological system is an obvious area for further research but has not been heuristic in leading to the discovery of any therapeutic agent or diagnostic test.

Side Effects

The original tricyclic antidepressant drugs had prominent anticholinergic side effects including dry mouth, GI constipation and blurring of vision due to blockade of the parasympathetic innervation of the eye. This biochemical property and these resulting side effects have been almost entirely eliminated in modern antidepressant monoamine reuptake blockers. However, some psychiatrists with the theoretical support of the Janowsky hypothesis of cholinergic-monoaminergic balance in the etiology of depression, have postulated that removal of anticholinergic properties has actually reduced the efficacy of modern antidepressants. Sometimes a trial of imipramine is clinically worthwhile in an SSRI unresponsive patient. Imipramine was not one of the switch options in the STAR-D trials.

Imipramine and other early tricyclic antidepressants carried potentially life-threatening side effects on overdose, especially overdoses with suicidal intent. These included cardiac arrythmias. These dangers have been eliminated in modern antidepressant drugs both SSRI and SNRI and these modern drugs carry a very low risk for successful suicide.

Reduction of sexual drive and increased time to achieve orgasm both in men and women are paradoxically the most troublesome side effects today in the illness of depression where loss of sexual desire is a cardinal symptom. Switching from an SSRI to an SNRI is often helpful. If the patient is a responder to SSRI and does not want to risk switching, sildenafil addition is sometimes helpful in male patients. Interestingly, SSRI are useful in the treatment of premature ejaculation: it is an ill wind that blows no man good (see chapter on sexual psychopharmacology below).

Pregnancy and Lactation

Many antidepressants have been associated with birth defects when used in pregnancy but these are difficult to define statistically because of the background prevalence of birth defects in untreated populations and also the possible association of birth defects with depression itself or comorbid substance abuse. If an antidepressant is considered necessary in a woman of childbearing age without contraceptive protection or a woman contemplating pregnancy while under treatment with an antidepressant, specific up-to-date literature on the specific antidepressant should be sought. Some antidepressants currently on the market seem to be without significant risk of physical birth defect. Similarly, many antidepressants are secreted in breast milk and should be avoided in lactating women or the mother should be advised not to breastfeed. Other antidepressants are not secreted in breast milk and can be consistent with lactation. The literature on the specific antidepressant must be consulted.

However, as with all drugs that enter the brain, the developing brain is a complex milieu where receptors and neurotransmitters are developing in concert with each other. It seems intuitive that blockade of monoamine reuptake in the developing fetus or in the newborn by antidepressant treatment would lead to changes in the laying down of monoaminergic brain circuits. Animal studies often suggest that treatment of a pregnant rodent with antidepressant compound affects the behavior of the offspring throughout their life. Of course, some of the changes reported in rodent research are difficult to extrapolate to human behavior and many not even be conclusively harmful: For instance, what if prenatal exposure to antidepressants led to a reduced fear response in the adult? Would this be bad or good? It is of course best, if possible, for pregnant women to avoid antidepressant treatment as well as all other drugs including alcohol, cannabis, nicotine, and even caffeine [17].

Animal Models and Depression Psychopharmacology

It is usual for the psychiatrist to ask whether depression can be modelled at all in animals. Clearly, no perfect model can exist since we do not know the etiology or even the true definition of depression in humans [18]. The Porsolt swim test puts a rodent in a swim bath from which he cannot escape, and after a few minutes he despairs and rests floating with his snout outside the water. Antidepressants in a single dose before the test increase the struggling time, the "persistence" of the rodent, and decrease "despair" time. This clearly does not mimic the course of clinical drug response in humans, which takes 3 weeks; but some reports of the Porsolt in humans (!!!) suggest that some psychological effects of antidepressants in humans also occur with the first dose. The Porsolt shows excellent predictive validity for the development of new antidepressants but alas all the new drugs seems to be merely "me-too variations" of the older ones. More complex psychosocial models of depression in animals, such as learned helplessness or maternal deprivation, show cortisol elevations but have not lead to any new pharmacological therapies. The monoamine systems and the cortisol systems interact in the brain, but we have not found a pharmacological way to utilize these interactions therapeutically. New research could be promising.

Stimulants in Depression

Amphetamines and similar substances such as methylphenidate and cocaine release neurotransmitters directly from the presynaptic neuron (see Fig. 5.5) including noradrenaline, serotonin and dopamine. Cocaine may also inhibit reuptake. All these substance cause an increase in activity in most (but not all) adult humans, a sense of good mood and even elation, increased sexual interest, increased self confidence, increased aggression, loss of need for sleep, and increased sociability. They work almost immediately and their effect lasts several hours to a day. Are they useful in depression? In almost all patients, no. Depressed people usually experience anxiety and dysphoria with them and they can not be taken repeatedly because of tolerance and addiction. More about them in the chapter on stimulant use in ADHD in children. Their lack of usefulness in clinical depression was one of the reasons for a revision of the monoamine theory of depression: the revision postulated that decreased adrenergic receptor sensitivity that requires about 3 weeks to develop was the real reason behind the monoamine receptor reuptake drugs slowly developing therapeutic effect [19]. But if so, why does not propranol that blocks adrenergic receptors immediately cause an antidepressant effect? The truth is that there are many contradictory pieces of the biochemistry puzzle of human antidepressant action and summaries with simple graphs and theories are just not the way forward for clinical thinking or research progress. A major difference between the antidepressants and the stimulants is that animals will work to get stimulants and humans get addicted: no animals will press a lever to get an antidepressant and no human raises his or her antidepressant dose repeatedly for the pleasure it gives.

An exception to the above are some geriatric depressions with anhedonia, motor retardation and lack of energy as chief symptoms. A morning dose of a stimulant such as methylphenidate 10 mg daily can sometimes be very helpful. Parkinson's Disease and hypertension should be ruled out [20, 21].

Antidepressants in Depression After Myocardial Infarction

While the thrust of the data today and our discussion above is that antidepressants are effective only when the depth of symptoms and absence of psychosocial precipitating causes suggest a "biological depression", there are notable exceptions. One exception is the deep and often hopeless depression that can follow a myocardial infarction Many studies find antidepressants helpful [22].

Antidepressants and COVID-19

The rapidly evolving literature on mood disorder risk for COVID 19 and possible risk of antidepressant use and COVID mortality must be a constant source of update for every practicing psychiatrist [23].

Psychological Effects of Antidepressant Treatment

Kramer, in his famous book "Listening to Prozac" [24], reported that patients whom he was following in psychotherapy reported changes in their dreams, their attitudes and their psychodynamics after starting Prozac treatment for depression without even being conscious that they were having an antidepressant response. His book strongly supported the emerging popular idea that Prozac allows people to feel better in the full sense of the word and has salutary effects on the human personality and life's many woes. Research efforts to confirm Kramer's thesis have been unsuccessful although it is hard to study normal volunteers on any long term psychotropic medication. Some studies have reported that the fear response of the human amygdala studied with functional MRI or PET scan in normal volunteers or depressed patients is altered by administration of SSRI and this might support Kramer's position [25, 26]. Animal studies have indeed revealed many effects of monoamine reuptake inhibitors on the stress response and behavior of animal models. However, it is not certain from the animal models if these effects are always positive. In recent years there has been growing criticism of the effects of antidepressant drugs on the human personality in a direction opposite to those suggested by Kramer: Some patients report that their emotional life has been blunted since beginning antidepressants [27]. Often patients report difficulties in stopping antidepressants and claim to feel a return of anxiety and depression greater than that experienced before beginning the treatment. These rebound effects or emotional blunting effects have not been demonstrated in a scientifically convincing way. However, the large number of effects of these drugs in normal rodent and primate brains suggests that they are not "magic bullets" for depression but may well have other effects on human personality and even on the social interactions of individuals with his or her family and place of work. Often researchers of the possible side effects of antidepressants in humans are not well versed in the animal studies and vice versa. There is an urgent need for research on the effects of these drugs on human lives other than their effects as antidepressants since such a large fraction of the population is being exposed to them today [28].

Clinical Vignettes

1. Ofer was a 40 year old married colonel in the military with a successful career and no previous emotional problems. He was promoted and transferred to a command at a new base. Within 3 weeks he began sleeplessness, crying and by a month he had lost six kilos of weight and developed psychomotor retardation and suicidal ideation. On admission to the day hospital he was diagnosed as having an endogenous depression and begun on an antidepressant. Over the next few days and in a circuitous manner details of his situation emerged: his career success was due to a talented female administrative secretary with whom he had had a romantic relationship over the course of several years. This relationship ended with his promotion and transfer and he found himself unable to cope without her help. Psychotherapy was begun, antidepressants were stopped and he was transferred to a less demanding military post with remission of his depression, now diagnosed as situational or reactive.

2. David was a 6o year old jewelry store owner on the main street in a small town with no previous history of emotional disturbance. He was hospitalized at the inpatient psychiatry unit with crying spells and sleeplessness but the most prominent symptom was a constant repetition of the words "all is lost, all is lost" explaining that his shop had gone bankrupt and therefore his family would starve and he would not be able to support his children through school. A diagnosis of reactive depression was made to the obvious financial catastrophe and he was scheduled for intensive group therapy sessions. Over the next few days his family met with a treatment team and explained that the shop was thriving, there was no financial crisis at all and they were all well cared for economically. The depression, they said, was completely out of the blue. The diagnosis was changed to endogenous depression and he responded within a month to antidepressants.

3. Sean was a shy 23 year man who went out on his first date and was so anxious that he vomited on her over dinner. He then developed recurrent thoughts about the event along with typical panic attacks several times a week. Because these happened more outside his home, he began staying at home and had not been outside it for 6 months when he came to the Psychiatry Outpatient Clinic. The psychiatrist had just read Donald Klein's 1964 article "Delineation of two drug responsive anxiety syndromes" [15] and prescribed low dose imipramine. Within days the patient stopped having panic attacks and with mild encouragement began leaving his home. He made a full recovery on last follow-up.

4. Ben, a family man in his 30s, had been treated pharmacologically for dysthymia the past 10 years, receiving noradrenergic and serotonergic blockers at full doses, for adequate periods of time, without significant recovery. Therapy for him had become the continuing search for the drug which would heal him. While clearly dysphoric, his suffering was not episodic, nor did he have prominent vegetative symptoms. After all the failed attempts, and in light of the nature of his depression, the psychiatrist suggested stopping medication, which Ben warily agreed to do. A closer examination of his distress uncovered environmental stressors related to his family, and personality issues of unresolved narcissistic injuries. Mindfulness exercises helped him get better control of the downward spirals. He responded well to insight-oriented therapy. No medication was reinstituted.

References

1. Klein D, Davis J. Diagnosis and drug treatment of psychiatric disorders. 1st ed. Williams & Wilkins; 1969.
2. Dean CE. The skeptical professional's guide to psychiatry: on the risks and benefits of antipsychotics, antidepressants, psychiatric diagnoses, and Neuromania. 1st ed. Routlage; 2020.
3. Geddes JR, Carney SM, Davies C, Furukawa TA, Kupfer DJ, Frank E, et al. Relapse prevention with antidepressant drug treatment in depressive disorders: a systematic review. Lancet. 2003;361(9358):653–61.
4. Machmutow K, Meister R, Jansen A, Kriston L, Watzke B, Härter MC, et al. Comparative effectiveness of continuation and maintenance treatments for persistent depressive disorder in adults. Cochrane Database Syst Rev. 2019;5(5):Cd012855.
5. Healy D. The antidepressant era. Harvard University Press; 1999.
6. Horwitz AV, Wakefield JC. The loss of sadness: how psychiatry transformed Normal sorrow into depressive disorder. Oxford University Press; 2007.
7. Kirsch I, Deacon BJ, Huedo-Medina TB, Scoboria A, Moore TJ, Johnson BT. Initial severity and antidepressant benefits: a meta-analysis of data submitted to the Food and Drug Administration. PLoS Med. 2008;5(2):e45.
8. Turner EH, Matthews AM, Linardatos E, Tell RA, Rosenthal R. Selective publication of antidepressant trials and its influence on apparent efficacy. N Engl J Med. 2008;358(3):252–60.
9. Belmaker RH, Agam G. Major depressive disorder. N Engl J Med. 2008;358(1):55–68.
10. Cipriani A, Furukawa TA, Salanti G, Chaimani A, Atkinson LZ, Ogawa Y, et al. Comparative efficacy and acceptability of 21 antidepressant drugs for the acute treatment of adults with major depressive disorder: a systematic review and network meta-analysis. Lancet. 2018;391(10128):1357–66.
11. Patel V. Scale up task-sharing of psychological therapies. Lancet. 2021;399:343–5.
12. Lichtenberg P, Belmaker RH. Subtyping major depressive disorder. Psychother Psychosom. 2010;79(3):131–5.
13. Stewart JW, Tricamo E, McGrath PJ, Quitkin FM. Prophylactic efficacy of phenelzine and imipramine in chronic atypical depression: likelihood of recurrence on discontinuation after 6 months' remission. Am J Psychiatry. 1997;154(1):31–6.
14. Belmaker RH. Treatment of bipolar depression. N Engl J Med. 2007;356(17):1771–3.
15. Klein DF. Delineation of two drug-responsive anxiety syndromes. Psychopharmacologia. 1964;5:397–408.

16. Rush AJ, Trivedi MH, Wisniewski SR, Nierenberg AA, Stewart JW, Warden D, et al. Acute and longer-term outcomes in depressed outpatients requiring one or several treatment steps: a STAR*D report. Am J Psychiatry. 2006;163(11):1905–17.
17. McAllister-Williams RH, Baldwin DS, Cantwell R, Easter A, Gilvarry E, Glover V, et al. British Association for Psychopharmacology consensus guidance on the use of psychotropic medication preconception, in pregnancy and postpartum 2017. J Psychopharmacol. 2017;31(5):519–52.
18. Kara NZ, Stukalin Y, Einat H. Revisiting the validity of the mouse forced swim test: systematic review and meta-analysis of the effects of prototypic antidepressants. Neurosci Biobehav Rev. 2018;84:1–11.
19. Lerer B, Ebstein RP, Belmaker RH. Subsensitivity of human beta-adrenergic adenylate cyclase after salbutamol treatment of depression. Psychopharmacology. 1981;75(2):169–72.
20. Zohar J, Belmaker R. Treating resistant depression. New York: Spectrum Press; 1987.
21. Smith KR, Kahlon CH, Brown JN, Britt RB. Methylphenidate use in geriatric depression: a systematic review. Int J Geriatr Psychiatry. 2021;36(9):1304–12.
22. Kim JM, Stewart R, Lee YS, Lee HJ, Kim MC, Kim JW, et al. Effect of escitalopram vs placebo treatment for depression on long-term cardiac outcomes in patients with acute coronary syndrome: a randomized clinical trial. JAMA. 2018;320(4):350–8.
23. Vai B, Mazza MG, Delli Colli C, Foiselle M, Allen B, Benedetti F, et al. Mental disorders and risk of COVID-19-related mortality, hospitalisation, and intensive care unit admission: a systematic review and meta-analysis. Lancet Psychiatry. 2021;8(9):797–812.
24. Kramer PD. Listening to Prozac: the landmark book about antidepressants and the remaking of the self. Penguin Books; 1997.
25. Gorka SM, Young CB, Klumpp H, Kennedy AE, Francis J, Ajilore O, et al. Emotion-based brain mechanisms and predictors for SSRI and CBT treatment of anxiety and depression: a randomized trial. Neuropsychopharmacology. 2019;44(9):1639–48.
26. Selvaraj S, Walker C, Arnone D, Cao B, Faulkner P, Cowen PJ, et al. Effect of citalopram on emotion processing in humans: a combined 5-HT(1A) [(11)C]CUMI-101 PET and functional MRI study. Neuropsychopharmacology. 2018;43(3):655–64.
27. Goodwin GM, Price J, De Bodinat C, Laredo J. Emotional blunting with antidepressant treatments: a survey among depressed patients. J Affect Disord. 2017;221:31–5.
28. Pratt LA, Brody DJ, Gu Q. Antidepressant use among persons aged 12 and over: United States,2011–2014. NCHS Data Brief. 2017;283:1–8.

Chapter 6
Antipsychotic Drugs: Do They Define Schizophrenia or Do They Blunt All Emotions?

The history of the serendipitous discovery of the antipsychotic drugs in the early 1950s is well known as is the general historical consensus that the introduction of these drugs contributed to a large decrease, and in some countries abolition, of the huge mental hospitals where psychotic patients had lived out most of their lives before that time. The antipsychotic drugs have had numerous names in professional terminology including neuroleptics and major tranquilizers, but these different names reflect the theoretical background of the users at the time rather than any true increase of knowledge about their mode of action or therapeutic indications. When they were introduced, most patients in chronic psychiatric hospitals had a diagnosis of schizophrenia. Actually, a larger percentage of the acute admissions were for psychotic mania or psychotic depression, but these affective psychoses tended to remit within months and were discharged. Most, although not all, of the patients with schizophrenia remained for chronic hospitalization [1].

This situation that existed in the era before antipsychotic medication was discovered in the 1950s was not always the case. Before the discovery of penicillin in the 1940s, or perhaps more accurately before the discovery of malarial and other treatments of tertiary syphilis of the CNS, a large fraction of the patients in psychiatric hospitals were suffering from the last stage of syphilis [2]. Tertiary syphilis causes a psychosis with paranoid delusions and hallucinations and most historical sources report that it was indistinguishable to clinical psychiatrists from what later came to be called schizophrenic psychosis. The difference was that some of the psychotic patients eventually developed general paralysis (paresis) and were then described as suffering from general paresis of the insane. On autopsy their brains were found to show the classical guma of tertiary syphilis. Those psychotic patients that did not develop paresis had no brain lesions on autopsy and these chronic psychotic patients were called schizophrenic. After the discovery of the syphilitic treponeme that solved the mystery of the cause of general paresis of the insane, it is understandable that many psychiatrists expected schizophrenia to ultimately reveal its true organic cause [3].

R. H. Belmaker, P. Lichtenberg, *Psychopharmacology Reconsidered*, https://doi.org/10.1007/978-3-031-40371-2_6

The effect of chlorpromazine, the first antipsychotic, on chronically psychotic patients in numerous hospitals where it was first tried was remarkable and dramatic. Delusions and hallucinations improved and the early descriptions also reported that apathetic, anhedonic and avolitional states also improved. The results were so remarkable that in a previous era chlorpromazine might have become the standard of treatment without scientific controlled clinical trials. However, the post World War II era was characterized by a newfound skepticism about medical treatments and the championing of the controlled double blind clinical trial by Lasagna and other medical leaders. Another cause of the skepticism about the reports of antipsychotic drugs being effective in schizophrenia was the rise of Freudian influence and the psychoanalytic movement in the United States and elsewhere following World War II. The rise of the psychoanalytic movement and its influence in psychiatry had many causes including critical elements of truth within the theory and also the sense of optimism post World War II that human nature if understood could be adjusted to bring happiness via talking and community therapies.

Rating Scales

But how could one measure improvement in an illness such as schizophrenia where symptoms are so subjective, laboratory or x-ray tests do not exist, and symptomology varies so greatly from patient to patient? The answer was the development of the psychiatric rating scale. The original one used in most early trials, the Brief Psychiatric Rating Scale (BPRS), is still widely used today. It originally used 16 items illustrated in Table 6.1. The rating scale is completed by a clinician who may be a psychiatrist or a psychologist but could also be a trained interviewer with some experience and talent for mental health work. The interview could be an open-ended interview lasting 15 min or a standardized clinical interview lasting 1–2 h. On each item the patient is rated 1 (not-present) to seven (extremely severe). Of course, the accuracy of such a rating depends on the experience of the interviewer. By summing up the rating noted for each item, a total BPRS score is obtained.

If this is a rating scale used for patients with schizophrenia, why are symptoms such as anxiety, depression and hostility included in the total rating? This has been discussed at length in psychometric research but in brief one could summarize that anxiety in schizophrenic patients often reflects frightening hallucinations or delusions, just as depression in such patients or aggression in these patients often reflect depressing or threatening hallucinations and delusions. It must be noted carefully that the BPRS never was used to diagnose schizophrenia and is not used in any system of diagnosing schizophrenia. It is a scale developed to reflect as accurately as possible the improvement seen by early clinical observers in patients diagnosed with schizophrenia and treated with the newly discovered antipsychotic drugs. Clinical trials were organized internationally based on the same model used in other areas of medicine at the time: Patients were randomly assigned to receive active chlorpromazine or an identical placebo tablet. They were rated on the BPRS at a

Table 6.1 Items in brief psychiatric rating scale

		1	2	3	4	5	6	7
1.	Somatic concern							
2.	Anxiety							
3.	Emotional withdrawal							
4.	Conceptual disorganization							
5.	Guilt feelings							
6.	Tension							
7.	Mannerisms and posturing							
8.	Grandiosity							
9.	Depressive mood							
10.	Hostility							
11.	Suspiciousness							
12.	Hallucinatory behavior							
13.	Motor retardation							
14.	Uncooperativeness							
15.	Unusual thought content							
16.	Blunted affect							

Fig. 6.1 A schematic example of the effect of an antipsychotic drug (CPZ = chlorpromazine) or placebo on the Brief Psychiatric Rating Scale s over 4 weeks

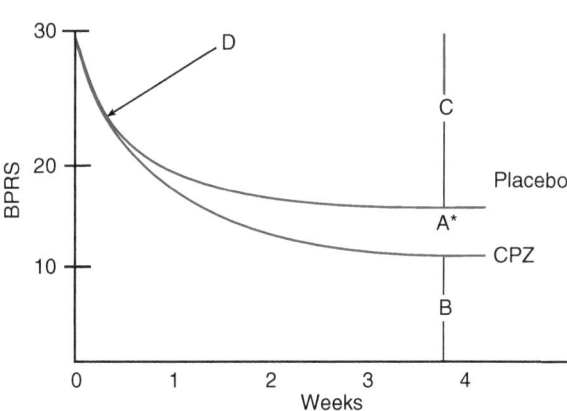

specific time point on beginning the study and then weekly after that perhaps for six or eight weeks. The results were dramatic and the archetypical result can is illustrated in Fig. 6.1:

Pharmacodynamics

The results of Fig. 6.1 have been repeated numerous times and have four critical components.

Point A: Significant differences between the placebo group and the chlorpromazine treated group begin at week 3. Kapur [4] has written that a proper statistical analysis shows that some symptoms and some patients do improve much earlier, and this clinical point is very important in guiding biochemical researchers to look for mechanisms of action that begin within the first few doses (such as dopamine receptor blockade) versus those biochemical effects that require chronic treatment (such as changes in gene transcription).

Point B: Even after chlorpromazine has reached its maximal effect after 3 or 4 weeks, considerable symptomatology remains or a considerable number of patients remain unimproved. Many statistical analyses of antipsychotic treatment have found that both are true: some symptoms remain in at least half of successfully treated patients and some patients, perhaps 10%, are completely unresponsive in their acute psychosis to dopamine blocking treatment [5].

Point C: Many patients or many symptoms improve in the placebo group. Studies have compared placebo tablets to phenobarbital; non-specific sedation does not account for the improvement in the placebo group over time in acute psychosis [6]. All inpatients in pharmacological trials of this kind are receiving good nutrition, protection from conflicts with family or neighbors, treatment of concurrent medical conditions that may have been neglected, and nursing and psychological treatment. Thus, the placebo group response represents the total response to all other treatments other than the compound under study, and should not be considered a response to suggestion or expectation alone. The passage of time also heals for many and reflects the course of illness: An example would be the asthmatic patient who arrives in the emergency with difficulty breathing: Often a dark room, guarantee of medical intervention if necessary, and reassurance allow the asthma attack to pass and any pharmacologic treatment must be compared to this placebo response. This use of the term "placebo" must be distinguished from studies, usually in pain, that find short term clinical benefit and neurophysiological correlates of the act of taking a pill [7].

Point D: There is a strong tendency for both the placebo group and active treatment group to improve by week one and perhaps before if a rating scale was done earlier but this improvement is common to both groups and does not represent a specific effect of chlorpromazine [8].

These classical results are a matter of little dispute today. The issues that have come to the forefront in recent years and will be discussed below are whether this treatment is specific in any way to schizophrenic psychosis, whether it is specific to psychosis or occurs in any disturbed mental condition, whether it reflects a specific biological or biochemical preexisting imbalance in these patients, whether it justifies continued and long term treatment of these patients and whether these results show that all patients who meet some set of diagnostic criteria should be treated with these drugs or whether some groups can improve without medication and whether some groups have so little benefit from medication that they might as well remain without the side effects [9].

Within a short time after the early acute studies of chlorpromazine and other antipsychotic drugs, the issue was raised as to whether patients who had been stabilized and perhaps discharged from hospital on these drugs need to be maintained continually on the medication or whether they had in any sense been "cured" and can go off the medication. A slightly different design was developed for this purpose and an archetypal study is shown in Fig. 6.2. Patients were admitted to study when

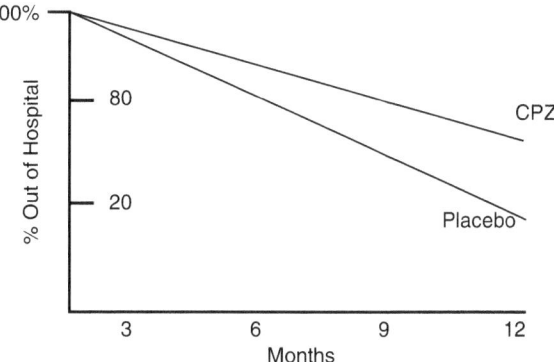

Fig. 6.2 A schematic prevention of relapse trial after hospital discharge comparing antipsychotic (CPZ = chlorpromazine) with placebo

they had reached remission by some definition or perhaps had been released from hospital. They were then randomized as in the acute studies to chlorpromazine tablets or placebo, identical looking tablets with inert contents. Patients were usually followed not with a rating scale such as BPRS but measured with a non-continuous variable such as need for readmission to hospital or relapse. Most studies lasted a year and some lasted 2 years. Typically, chlorpromazine treated patients had a relapse rate of about 30% by the end of 1 year whereas placebo treated patients had an 80% relapse rate as can be seen in Fig. 6.2.

The Disease Model

These clinical studies that antipsychotic drugs were effective in both the treatment and maintenance prevention of hospital relapse astounded the psychiatric world of the 1950s and have remained true to this day [1, 2]. However, it has been convenient for us in psychiatry as in many other areas of medicine, to avoid asking some hard questions which are now being asked:

1. What about those one third of patients who seem to respond to placebo? Is it ethical and good medicine to expose them to the side effects of antipsychotic drugs [10]?
2. What about the one third of patients who do not respond to or respond only minimally to antipsychotic drugs? Is it ethical and good medicine to keep them on antipsychotic drugs or would it be better to manage their psychosis in some other way?
3. Are these drugs antipsychotic in the same was that penicillin is an antibiotic that attacks foreign organisms with cell walls whose manufacture is inhibited by the penicillin molecule and whose biochemistry is so different from that of mammalian cells that there are few direct side effects? Or do antipsychotic agents affect other areas of brain functioning and change the nature of the patient's experience of his reality and self in a way that creates a high price for the benefits that he receives?

4. Is there any rebound of psychosis if antipsychotic drugs are stopped after successful treatment such that the patient becomes dependent on continued administration of these drugs?
5. Does the fact that these drugs also work in mania and in severe depression affect our conceptualization of them as specific for psychosis?
6. Should these drugs be considered "indicated" in schizophrenia in a DSM or ICD based diagnostic framework as has been increasingly the case in the last 70 years, or should they be conceived of as a clinical tool for the symptom of psychosis in many psychiatric situations in a way that emphasizes the need for multiple other treatments rather than using the antipsychotics as if they were magic bullets?
7. Can these drugs really be used safely in the developing brain of children for conditions as widely disparate as aggressive behavior or "temper tantrums"?

Each of the above questions are at the forefront of the thinking of clinical psychiatrists around the world today, but their open debate has often been obstructed by the needs that the profession has developed for a medically organized system of diagnosis. One classical model of medicine taught us that one must diagnose before treatment and then give the treatment on the basis of diagnosis. This model is useful when the surgeon must distinguish between acute appendicitis and acute gall balder disease, because it would not be helpful to take out the wrong one of those two organs in the case of acute abdomen. However, in many areas of medicine the diagnosis often is not a clear indicator of treatment. Chronic obstructive pulmonary disease may respond to anti-allergic treatments even if its main causal factor in a particular patient is smoking. Hypertension may respond to beta adrenergic blockade even if the etiology in a particular patient may be salt intake. Clearly syndromes such as pain are treated with a variety of medications that affect different receptors and pathways often unrelated to the cause of the pain [11].

Mechanisms of Action

Before we discuss each of the above limitations in turn, it is worth discussing some more of the historical events in the development of the antipsychotic drugs. The tremendous success of chlorpromazine in the treatment and prophylaxis of psychotic episodes of schizophrenia or bipolar disorder led to its chemical imitation by competing pharmaceutical companies. There was great hope for many years that each new chemical permutation of the chlorpromazine molecule might lead even accidentally to a better and more effective antipsychotic. A long list of drugs was developed including; perphenazine, trifluoperazine, thioridazine, and fluphenazine.

Each new one was heavily advertised and often claimed on the basis of case reports or key opinion leaders to be unique in one or another subtype of schizophrenia. None of those studies have replicated consistently and by the 1970s it became clear that all of these antipsychotic drugs at the appropriate dose had equivalent clinical efficacy [12]. The appropriate dose was decided by comparing doses in large groups of patients with the endpoint being remission from acute psychosis and, less frequently, prevention of relapse to hospitalization as a gross measure. Tables of recommended doses were published in psychiatric textbooks. The realization that these drugs were clinically equivalent led to the search for a common mechanism of biochemical action. The initial studies of these drugs on the biochemistry of the brain was limited by the knowledge at the time. Interestingly, the search for a mechanism of action of the antipsychotic drugs was one of the key heuristic drivers of the explosion of knowledge of brain neurochemistry described in Chap. 2. The antipsychotic drugs were found to inhibit cholinergic receptors, to inhibit alpha adrenergic receptors, to inhibit dopamine receptors and to block serotonin receptors. In addition, they stabilized lipid membranes, affected the nerve action potential and affected the levels of hormones such as prolactin in the peripheral blood. Which of these many effects could possibly be related to the mechanism of action? A breakthrough idea was the realization that the different clinically equivalent antipsychotic drugs each possessed only some of these chemical properties.

Only dopamine receptor blockade was a common element to all of the clinically effective antipsychotic drugs. Moreover, using John Davis's table of optimum clinical dosages and using the new technique of radioactive labeling of a dopamine receptor ligand, Sol Snyder [13] could show that all antipsychotic drugs were dopamine receptor blockers with an efficacy proportional to the clinical dose useful to optimally help acute psychotic patients (see Fig. 6.3).

This astounding correlation seemed to be as clear a proof as one could come to showing that dopamine receptor blockade was the mechanism of action of antipsychotic drugs. That is, if one gives a patient who has begun to believe that voices are telling him that he is the messiah, that microphones have been installed in his walls to follow his actions by the enemies of the messiah and that his former friends have really turned against him; and one gives such a person a chemical compound of a known composition and he begins to lose these symptoms over the coming weeks— one can know that the compound that one has given is affecting these changes in his

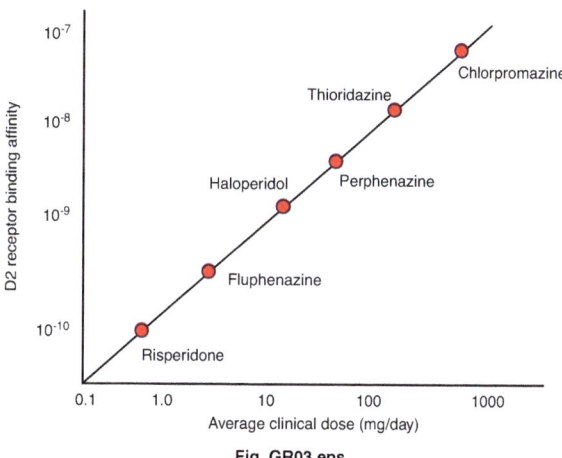

Fig. 6.3 Effectiveness of neuroleptics is highly correlated with binding affinity to DA receptors

Fig. GR03.eps

emotional, perceptual and behavioral state by binding and blocking his dopamine receptors. Most recently, the dopamine receptor genes have been sequenced, their amino acid structure determined, and their form in space and placement in the cell membrane so well characterized—that it is now possible in silico, in a three-dimensional computer model to design chemical compounds that bind to the receptor and that will inevitably turn out to be antipsychotic in the clinic. This indeed has been a spinetingling revelation for the last generation of psychiatry. It is still a meaningful and satisfying clinical experience that every clinician in the field has experienced. However, (see below) it does not negate the new questions that we have asked above.

The Dopamine Hypothesis

Sol Snyder [13] summarized that every compound that blocks dopamine receptors is antipsychotic, that every compound that is antipsychotic blocks dopamine receptors, and that every compound that blocks dopamine receptors and is antipsychotic does so with a clinically optimal dose that is well correlated with the in vitro dose necessary to block the dopamine receptor. This finding became one of the foundation stones for the dopamine hypothesis of schizophrenia. Another foundation stone for the dopamine hypothesis is the observations beginning in the 1950s of the high rate of psychoses in amphetamine abusers or abusers of similar stimulants such as methamphetamine and cocaine. These drugs when used at high enough doses for long enough time produce a psychosis often indistinguishable from acute psychosis. Moreover, the psychoses are treated effectively with antipsychotic drugs and the symptoms of psychosis often remit within hours. These drugs such as amphetamine and methamphetamine and cocaine were found in laboratory rodent and primate studies to be releasers of dopamine at the dopamine synapse and thus increase dopamine function in the relevant brain areas. These findings seem to complement the finding of the role of antipsychotic drugs in the treatment of acute psychosis. Amphetamine or related dopamine releasers or dopamine receptor agonists cause behavioral changes in rats or monkeys that are eerily reminiscent of psychosis: a rat might repeatedly scratch a fictive itch, chew a nonexistent food pellet, and a monkey might turn its head to listen to a nonexistent voice and search repeatedly for a nonexistent enemy. Perhaps, reasoned leaders of psychopharmacology, schizophrenia is an illness of excess dopamine that is treated by blockers of dopamine receptors? This hypothesis called the dopamine hypothesis of schizophrenia has been a major area of schizophrenia research for the last 50 years.

The dopamine hypothesis of schizophrenia was examined with a wealth of research technology and effort since its formulation in the 1970s. Early studies used methods such as measuring the concentration of HVA (homovanillic acid), the major metabolite of dopamine, in the spinal fluid of patients with schizophrenia vs. controls. When collections of properly stored post-mortem brain of patients diagnosed with schizophrenia became available, studies measured the number of receptor binding sites for dopamine in schizophrenia patients vs. controls. When ethically possible in some centers, researchers recruited volunteers to observe the effect of amphetamine to exacerbate the symptoms of schizophrenia in remitted patients vs. controls. Recent studies use PET (positron electron tomography) to measure the release of dopamine in the brains of living patients with psychosis compared with controls. These various technologies have been used heuristically and energetically by a generation of psychiatric researchers. One of the authors (RHB) studied mono-amine oxidase enzyme activity in platelets of patients with schizophrenia, reasoning that low MAO might be a risk factor for enhanced dopamine if it could not be ade-quately metabolized. Low levels were found [14]. However, after hundreds of attempts at replication and expansion of this finding in many laboratories and many encouraging as well as discouraging data, it was concluded by this author (RHB) that platelet MAO can be affected by platelet count, by diet, by exercise and by stress and that low platelet MAO is more likely a byproduct of mental illness than a biomarker or cause [15].

Psychiatry can be justly proud of the mobilization to use a testable hypothesis formulated as a result of the discovery of antipsychotic drugs. However, these efforts have not generated a biochemical pathophysiology of schizophrenia. Early studies such as the levels of HVA in the spinal fluid of schizophrenic patients often found positive results, later refuted by negative results and often summarized in a review article and in later years with a meta-analysis. As new technologies devel-oped, the old findings quietly disappeared from the literature rather than ever being refuted entirely. Thus, an ahistorical reading of the literature has often led a trusting student to believe that many abnormal findings relating to dopamine in the schizo-phrenic patient exist. However, if such a student or patient were to attempt to find a hospital to make that measurement so as to verify a diagnosis of schizophrenia, he or she would be met with bewildered confusion. No biochemical findings have been replicated and verified that separate adequately in a diagnostically useful way between groups of patients who clinically have schizophrenia and controls. Even if this were to happen, for diagnostic use we would face the even higher bar of requir-ing that the dopamine abnormality be present only in schizophrenia by comparing schizophrenic patients not only with normal controls but with patients in mania, psychotic depression or perhaps in other states of heightened aggression.

New Antipsychotics

As the number of dopamine blocking variations of the phenothiazine molecule increased, other molecules were found that could also block dopamine receptors. Haloperidol was one of the first non-phenothiazine antipsychotics and was chemi-cally a butyrophenone. It was discovered by screening chemicals for ability to block behavioral effects of amphetamine in the laboratory. While each new compound was greeted historically with enthusiasm and often major commercial success for the pharmaceutical industry, eventually the consensus has been reached that none of these dopamine blocking antipsychotics had any specific spectrum of clinal effects and none were superior to the other. There was some variety of their side effects, which is disused below. Quite serendipitously, a newly developed antipsychotic called clozapine began clinal trials without particularly high expectations by the developers in the pharmaceutical industry. Initial clinical trials reports (later verified as true), that it occasionally causes agranulocytosis and can be fatal, led to its withdrawal from the market. In an unprecedented fashion, families of patients who had participated in the early clinical trials organized protests out of their strong conviction that clozapine had given their loved ones unique clinical benefit. These protests were heard by several psychiatrists, who can be considered heroes of the field, including Nathan Kline and John Kane and in the late 1980s published their controlled clinical trial of clozapine in severely ill schizophrenic patients unresponsive to other treatments [16]. This classic study has been generally replicated and clozapine is accepted as having unique properties for some patients with schizophrenia and its serious risk of agranulocytosis is managed clinically by a strict regime of frequent monitoring of white blood count, weekly for at least 6 months, and assurance that a nearby hospital is equipped with supplies of CSF (colony stimulating factor) for treatment of any patient that may develop agranulocytosis due to clozapine. Clozapine is still worth the risk for a significant group of patients unresponsive to other therapies.

The serendipitous discovery of clozapine led to a wonderful race by pharmaceutical companies eager to find a clozapine like medication without the agranulocytosis. At first this seemed simple: Find the part of the clozapine molecule that looks toxic to an expert in medicinal toxicology and produce a clozapine without that offending site. This didn't work so other ideas came into play. Clozapine seemed to have unique inhibitory properties on the dopamine D4 receptor, since by the 1990s molecular biology had shown that the natural neurotransmitter dopamine had five different receptors coded for by five different genes and with somewhat different pharmacological properties and anatomical placement in the brain [17]. Perhaps specific dopamine D4 receptor blockers would be clinically beneficial like clozapine? This turned out not to the case and such pure D4 blockers are probably not even antipsychotic at all. Another hypothesis relates to the fact that clozapine inhibits some of the serotonin receptors and well as the dopamine receptors. One of the first dopamine-serotonin receptor blockers was risperidone, which was widely marketed with great enthusiasm as possibly being more effective than the first generation of antipsychotics. Risperidone has fewer extrapyramidal side effects than most first generation antipsychotics (see below discussion of side effects of the antipsychotic drugs). This may be because a serotonergic system exists in balance with the dopamine system in the parts of the brain controlling posture and movement in the basal ganglia. However, no added benefit to resistant patients for risperidone has been shown as well as well as no added efficacy in general for psychotic patients. Another line of research was to take the clozapine molecule and make a permutation of it as close to the original as possible. This line of thought led to olanzapine. Like risperidone, olanzapine is a dopamine receptor blocker with some serotonergic blocking effects and is no more effective than the first generation of dopamine blocking antipsychotics. Olanzapine, while having few EPS, became the subject of a shocking story because of possible pharmaceutical company negligence in reporting a whole new set of side effects revolving around weight gain, decreased glucose tolerance and hyperlipidemia. Lilly Pharmaceuticals settled out of court the law suit relating as to whether they responded early enough and appropriately enough to this new internal data. However, no one doubts the efficacy of olanzapine as an antipsychotic and its dearth of EPS side effects.

Olanzapine and risperidone introduce a whole new sequence of yearly developments of new antipsychotics called by many the "second generation" antipsychotics. Some count the second generation as beginning from clozapine but actually none since clozapine has demonstrated the uniqueness that clozapine does possess. The almost yearly introduction of new second generation antipsychotics has been the life blood of the pharmaceutical industry in this field but somewhat of an embarrassment for scientific psychiatry. Each new compound seems to conquer the prescription market after introduction and heavy advertising only to be found not very different from its predecessors after a year or two and replaced with a new compound with almost identical claims. Emphasis in recent years has been to find a compound that is similar to olanzapine and risperidone in lacking EPS but also to avoid all of the "metabolic" side effects of olanzapine and compounds similar to it that cause hyperlipidemia, decreased glucose tolerance and weight gain. Being "weight neutral" is a major goal of antipsychotic drug development at the present.

In many ways the competitive and optimistic approach of the world pharmaceutical industry to finding a new clozapine like compound represents the efficient functioning of scientific capitalism. However, the pricing of drugs, the heavy advertisement that sometimes included unethical rewards for high prescribing doctors and the complicity of the FDA in allowing new compounds to be marketed as probably unique and more effective when they had only been shown to be better than placebo—these have been a disappointing and frustrating aspect of our current system of drug development. It might be better for the FDA to demand that new compounds be shown to be better than existing ones or with fewer side effects by head-to-head of comparison rather than to continue to allow faddish marketing on the basis of placebo controlled trials in antipsychotics.

Beginning with the now classic CATIE study [18] many meta-analyses have found that second generation antipsychotics are no more effective than first generation antipsychotics and no one of the second generation antipsychotics is more effective than any of the others. Interestingly, but somewhat disappointingly, some meta-analyses disagree with other meta-analyses, perhaps finding a slight advantage for one antipsychotic over another but no consistent pattern for the clinician has emerged. It is telling that some meta-analyses found amisulpride, a first generation antipsychotic that acts entirely as a dopamine D2 receptor blocker, to be perhaps slightly more effective than other antipsychotics [12].

Recent creative ideas have emerged, such as the concept that some compounds may be partial antagonists and partial agonists at the dopamine receptor. This hypothesis suggested that schizophrenia may result from dopaminergic hyperactivity in the limbic system as far as its positive symptoms are concerned, but may involve dopaminergic hypoactivity in the cerebral cortex as a cause of its negative or deficit symptoms. Partial agonists/antagonists stimulate the dopamine receptor when levels of dopamine are low but inhibit the dopamine receptor when levels of dopamine are high. The molecular mechanism of this effect is understood at the chemical level and is well demonstrated in pharmacology. Aripiprazole was designed on this basis as a partial dopamine agonist/antagonist. It was hoped that it would help the positive symptoms of psychosis by inhibiting dopamine receptors in those areas of the brain where there may be hyperactivity and help the negative symptoms of psychosis by stimulating dopamine receptors in those areas where dopamine may be underactive. This was an excellent idea but has been disproven in clinical trials: Aripiprazole is no more effective than other antipsychotics and does not have a special profile for negative symptoms. It is some improvement over existing second generation antipsychotics because it is weight neutral and does not cause hyperglycemia or hyperlipidemia, but it has many clinal limitations including high rates of akathisia and a limited dose-response range. At higher doses it is probably a pure dopamine receptor blocker and more like the first generation antipsychotics, some of which are also weight neutral [7].

Another recent development has been the synthesis of dopamine D-3 receptor antagonists, such as cariprazine. The D-3 receptor is inhibited by clozapine, and animal studies suggest that the D-3 receptor may have unique behavioral functions. Cariprazine is now widely sold for psychosis, but it is telling that its effect on D-2 receptors is also strong. It is too early to know if it will have clozapine-like effects in patients unresponsive to D-2 blockers and if other compounds might be effective if they are pure D-3 receptor blockers [19].

The disappointment at the lack of rapid progress in true effectiveness of antipsychotic drugs as opposed to the apparently commercial perception of rapid replacement of old drugs by new ones as illustrated in Fig. 6.4 has led to considerable

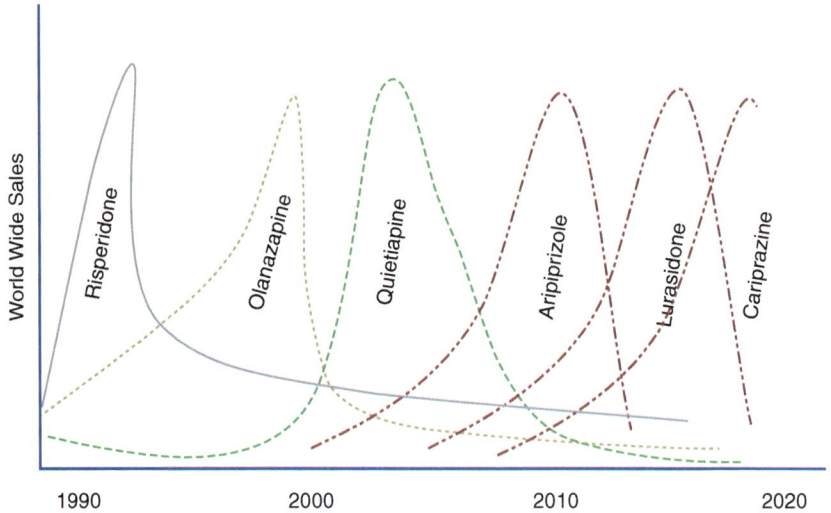

Fig. 6.4 A schematic artist's version of antipsychotic sales changes over time

skepticism about the nature and value of antipsychotics drugs, both among some patient advocacy groups and among psychiatrists themselves [10]. The authors of this volume share some of that skepticism but are clear in their belief that it is important not throw out the baby with the bathwater. Antipsychotics are a major tool in psychiatry but are not a cure all for schizophrenia. This would be the time to approach each of the seven points introduced above on page 8 as the hard questions that a textbook of psychopharmacology in 2023 must ask:

Philosophy of Antipsychotic Treatment

Point 1: What about those one third of patients who respond to placebo?

Patients in clinical practice with psychosis and who have had short or micro psychoses in the past that have resolved asymptomatically should not necessarily be put on antipsychotics, especially if they do not wish to be medicated. Such patients should be followed closely with stress reduction, psychotherapy and perhaps nutraceuticals such as omega-3 or folic acid that might have some pharmacological value as well as placebo value, even if the prodroma in such patients allows a DSM-5 or ICD-11 diagnosis of schizophrenia to be made. The clinician should avoid the knee-jerk response that DSM diagnosis of schizophrenia leads to an indication for antipsychotic treatment. Antipsychotic treatment has serious side effects and psychosis that can be suspected of having a benign course can sometimes be treated without antipsychotics [20].

Point 2: What about the one third of patients who do not respond to or respond only minimally to antipsychotic drugs?

Clinicians in public mental hospitals or clinicians who deal with homeless individuals often find chronically psychotic patients who have never responded to past treatment with antipsychotics and who are vehemently opposed to taking antipsychotic drugs. It is important that the clinician realize that the goal of psychiatry as a medical profession is not to give treatment but to maximize function and meaning of life as the individual sees it. A patient who does not benefit from antipsychotics and does not want the antipsychotics should be engaged to find other ways to maximize his goals for his own life. Psychopharmacology is not an end in itself but only one part of our treatment armamentarium.

Point 3: Are these drugs antipsychotic in the same way that penicillin is an antibiotic?

Antipsychotic drugs should better be named dopamine receptor blocking drugs which describes their function in the brain and does not connect them with a specific diagnosis such as schizophrenia or psychosis. There is no simple rocket launcher that can take psychopharmacology from the level of the synapse and blockade of dopamine to the level of behavior and improvement of psychotic behavior. There must be several explanatory stages in between. One famous attempt by Kapur [21] describes the key psychological mechanism called salience. Salience relates to the fact that numerous stimuli impinge on a perceiving human being at any given time.

A mechanism exists, perhaps in Freudian terms "the ego", that chooses actively among the multiple stimuli and gives attention to those stimuli that are relevant to the ongoing behavior. Thus, a patient speaking with his psychiatrist would be expected to give attention to what the psychiatrist is saying and to what he is reporting to the psychiatrist rather to the sound of a fly buzzing in the room or a dog barking outside the office. In psychosis the mechanism of creating and directing salience seems impaired. Kapur hypothesizes that antipsychotic dopamine blocking drugs reduce the salience of numerous irrelevant stimuli [21]. However, the cost of this reduction in salience when the patient becomes better is a reduced range of choice of his attention to different aspects of reality. Most patients who are beyond their acute psychotic phase dislike treatment with neuroleptics. Dopamine is well known to be involved in hedonic behavior and is often studied in the mechanism of addiction and self-medication with pleasure inducing drugs. It is no surprise that patients diagnosed with schizophrenia and other psychotic patients who receive long term treatment with antipsychotics complain that their ability to experience pleasure is reduced [22]. Psychiatric research has not given enough attention to the effects on subjective patient experience and to the price that the patient pays for antipsychotic treatment. One approach to this problem has been to reduce recommended doses of antipsychotic medication or to increase the frequency of follow-up such that antipsychotic medication can be eliminated for many patients and reintroduced rapidly if and when it is needed. A data based answer to this question is not yet possible, but several ongoing studies will give us answers in the near future as to whether and how we can reduce dosage of dopamine blocking drugs in acute psychosis and in prophylaxis [23].

Point 4: Is there any rebound of psychosis if antipsychotic drugs are stopped after successful treatment?

This question was a central concern in the early days of antipsychotic development and several studies found that no rebound occurs. Choinard et al. [24] reported a number of cases where patients seemed to have become dependent on their neuroleptic drugs in a behavioral sense so that stopping these medication led to a severe and malignant psychosis worse than the psychosis for which the patient was originally treated. Choinard's concept may well be true for a small minority of patients; it may also be the case for clozapine. It has not been proven to be the case for the majority of antipsychotic-treated patients. Whittaker [25] has developed this concept into an angry criticism of antipsychotic treatment in the twentieth century, essentially claiming that psychoses are benign and often self-limiting conditions and that modern antipsychotic treatment has caused an epidemic of dependence on dopamine blocking treatment that essentially makes psychosis worse when the treatment is stopped. Whittaker's claims are not backed by enough facts for them to make a coherent clinical policy. They run a significant danger of causing patients who do need antipsychotics to stop them [20]. However, they have been thought provoking and several studies of neuroleptic discontinuation are underway.

Point 5: Does the fact that these drugs also work in mania and in psychotic depression affect our conceptualization of them?

Antipsychotic drugs are highly effective in mania, both first generation antipsychotics and second generation antipsychotics. The classic textbooks of psychopharmacology that summarize the early literature such as Ban's Psychopharmacology in 1969 [1] or Klein and Davis "Diagnosis and Drug Treatment of Psychiatric Disorders" [2] in 1969 summarized the majority of studies of severe depression as finding chlorpromazine equal to imipramine. All studies find antipsychotics, both first generation such as haloperidol or chlorpromazine or second generation such as olanzapine or risperidone, to be at least as good as lithium in acute mania and in prophylaxis of bipolar disorder [26, 27].

Point 6: Should antipsychotic drugs be considered "indicated" in schizophrenia in a DSM-5 or ICD-11 based diagnostic framework?

Almost all studies of antipsychotic drugs used a DSM or ICD based definition of schizophrenia or psychosis or mania as part of the criteria for inclusion in the study, and then used the study as evidence that these drugs treat a disease as defined in the entrance criteria. This is clearly circular. The significant response to placebo, the significant rate of nonresponse to antipsychotics, and the significant effects of antipsychotics in illnesses such as mania and depression all suggest that dopamine receptor blockade is a symptomatic or syndromatic treatment and not a disease-based therapy.

Antipsychotic drugs in injectable form are a mainstay of pharmacological treatment of acute psychotic agitation or aggression and may be necessary to reduce the use of physical restraints and prevent harm to staff and other patients in inpatient settings or even to allow safe transport of acutely psychotic patients to hospital. There are few controlled studies of such patients because they cannot give informed consent and placebo control may not be an acceptable risk for caregivers. One of the authors (RHB) led a study of intravenous haloperidol vs intravenous diazepam in acute psychosis: both groups improved rapidly but equally, suggesting that such acute effects are sedative and unrelated to later antipsychotic effects [8]. Other clinicians have felt that acute antipsychotic treatment causes a "Parkinsonian straightjacket" of akinesia that may not be a therapeutic antipsychotic effect but is a justifiable medical intervention in some situations. Other contentious issues related to the specificity of antipsychotic drugs is their use in delirium (not indicated in delirium tremens treatable with benzodiazapines but sometimes helpful in in the ICU in other delirious states) and their widespread use in psychosis of dementia especially Alzheimer's with psychosis. Controlled trials find antipsychotics ineffective and perhaps dangerous in such organic psychoses; in practice their use appears unavoidable for behavioral control. The clinician should take care to lower the dose and discontinue antipsychotics when the acute need has passed in an Alzheimer patient.

Point 7. Can these drugs really be used safely in the developing brain of children for conditions as widely disparate as aggressive behavior or "temper tantrums"?

The use of antipsychotic drugs in children has exploded exponentially in recent years. The diagnoses are usually attention deficit disorder, disruptive behavior disorder, episodic temper dyscontrol disorder and sometimes childhood bipolar disorder. It is surprising that some clinicians conceive of the dopamine blocking drugs as somehow being specific to psychosis while on the other hand using them liberally in childhood aggressive behavior. The dopamine system is not fully developed in young children and certainly the cortical dopamine system, critical for the adolescent in developing abstract thought and impulse control, is not fully developed in children age 8–15. Blocking these receptors could conceivably have long term implications for the child involved. The theoretical conundrum of the use of antipsychotics in these disorders is of course overshadowed by the long-term iatrogenic dangers [28].

Critical to current thinking are studies of "high-risk adolescents" who have first degree relatives with psychosis and some symptoms of psychosis themselves and who can be categorized as a group with a 50% risk of developing psychosis within 24 months. These patients do not have reduced risk if treated with antipsychotics whereas some benefit may be achieved with antidepressants, omega-3 nutraceuticals or with psychosocial therapy [29]. The lack of benefit for early intervention with dopamine receptor blockers argues strongly against a disease model of schizophrenia involving dopamine.

Doses

Early clinical studies of antipsychotics such as chlorpromazine used the dose based on the phase II open clinical experience. Eventually, controlled clinical trials compared dosages such as chlorpromazine 200 mg daily vs. chlorpromazine 300 mg daily and found that higher doses up to about 800 mg a day suppressed psychotic symptoms most effectively in acute psychosis. Little attention was paid to side effects. Curiously, in retrospect, some leading psychopharmacologists regarded extrapyramidal parkinsonian side effects as a prerequisite for antipsychotic efficacy and a proof of adequate dosing. It was recognized that dosing for maintenance and prophylaxis of hospitalization was lower than doses for acute episodes but few comparative dosing studies were conducted in the maintenance phase. Little attention was given in these dosing studies to the subjective experience of the patient under high dose neuroleptic treatment and its effect on future compliance and personal satisfaction. One of the authors (RHB) with a colleague participated in a preparation for a clinical dosing experiment by taking 5 mg of haloperidol intravenously [30]. The two subjects experienced an avolitional state, with slowing of thinking and movement but severe inner restlessness for a couple of days. These troubling subjective experiences piqued some skepticism about the mystique that antipsychotic drugs are specific cures for psychosis.

Second generation antipsychotics with relatively few EPS have a whole range of metabolic side effects that did not even appear on the side effect rating scales used for first generation antipsychotics. Therefore, they appeared to be side-effect free! As their side effects became known a sense of dread and even shock spread through the psychiatric community. An epidemic of obesity and diabetes among individuals with serious mental illness has undoubtedly occurred which we are only now making major efforts to ameliorate. Part of that amelioration involves pharmaceutical company efforts to create weight neutral second-generation compounds, use of lower doses of antipsychotics, and use of antipsychotics only when necessary and for more restricted periods of time in our patients' lives. A psychotic diagnosis is not a proven lifetime indication for antipsychotic treatment and prophylaxis. The choice of a particular dose for a particular patient must include recognition of the patient's need for a subjective quality of life and the negative impact that antipsychotics often have on the subjective sense of well-being. This area of the proper dose for each patient in still part of the art of psychiatric medicine, and textbook tables of recommended dosages based on dopamine receptor binding affinity or consensus recommendations need to be recognized as weak evidence indeed.

With the advent of PET imaging, elegant studies showed that about 70% occupancy of the dopamine receptor in the basal ganglia reliably led to parkinsonian side effects. It was proposed that a specific lower level of dopamine receptor occupancy on PET could predict antipsychotic efficacy. This concept is probably outmoded since it is likely that the areas of the brain where dopamine receptor blockade reduces salience or otherwise leads to improvement of psychosis are elsewhere other than in the motor areas of the basal ganglia. The pharmaceutical companies tend to do their phase III clinical trials for FDA registration at doses large enough to prove efficacy but small enough to have minimal side effects. Dose recommendations on approval of a compound are often the beginning of a slide uphill of drug dosages in clinical practice. Often new side effects appear in Phase IV post marketing surveillance as with the occurrence of glucose intolerance and weight gain with olanzapine and other second-generation compounds. Clinicians must realize that dose reduction on follow-up of patients treated with antipsychotics, when possible, is as necessary as dose escalation when needed.

Side Effects

The most common side effects of first generation antipsychotics are extrapyramidal syndrome (EPS) or drug induced Parkinsonism. The symptoms include a typical parkinsonian tremor which when severe can lead to marked shaking and inability to even hold a cup; akinesia or Parkinsonian inability to initiate movement which when severe can lead to a *festinating* gait with small steps and a posture that appears to be falling forward and also mask facies, a lack of expression in the face that can mimic the apathy and emotional bluntness of schizophrenia; and akathisia, a paradoxical sense of inner restlessness that can be intensely uncomfortable and has even been considered one of the causes of suicide in newly treated patients [30]. These parkinsonian symptoms cannot be treated with L-dopa, the classic drug for Parkinson's, because L-dopa reverses the blockade of antipsychotic drugs and exacerbates psychosis. Instead, the anticholinergic compounds that are used to help Parkinson's Disease are also found to be very helpful in antipsychotic induced parkinsonism. This is explainable because of the balanced interaction of cholinergic and dopaminergic neurons in the basal ganglia (see Chap. 19). Blocking cholinergic receptors in the presence of blocked dopamine receptors restores a balanced input to the ancient motor systems that control posture and movement. About one third of patients treated with first generation antipsychotic treatment require treatment of EPS with anticholinergic drugs. A much smaller proportion of patients treated with second generation antipsychotics have this need but it should not be ignored if EPS is present. The anticholinergic anti-parkinsonism treatment comes with its own set of side effects because of presence of cholinergic receptors in every gland innervated in the human body by the parasympathetic nervous system. Thus, the treatment with anticholinergic medication for EPS side effects leads to dry mouth, gastrointestinal constipation, some tachycardia, dry skin, enlarged pupils and occasionally urinary retention. Anticholinergics should probably be used only in those patients who need them and not prophylactically together with onset of neuroleptic treatment. Akathisia, a more diagnostically challenging parkinsonian symptom, responds only occasionally to anticholinergic medication but does respond to propranolol or to benzodiazepines.

Another prominent side effect of first generation antipsychotics is hyperprolactinemia. The tubero-infundibular neurons in the hypothalamus secrete dopamine onto receptors in the pituitary gland which inhibit secretion of prolactin. Blockade of that inhibition leads to huge increases in circulating prolactin. Surprisingly, this only causes side effects in a portion of the patients although it is biochemically present in almost all neuroleptic first generation treated patients. In males the hyperprolactinemia can lead to decreased sexual desire and erectile dysfunction and sometimes to clinically significant gynecomastia. In female patients galactorrhea can be a worrisome side effect, particularly frightening to a patient who may have delusions about being pregnant. Often and perhaps even in a majority of females treated with first generation neuroleptics, menses cease. Treatment with bromocriptine, a dopamine agonist that penetrates the CNS poorly, is less dangerous than L-dopa. However, second generation neuroleptics present a better alternative for patients with these side effects from first generation treatment.

A rare side effect, even rarer in second generation antipsychotics, is neuroleptic malignant syndrome (NMS), including hyperthermia, muscle rigidity, increased serum CPK, clouding of consciousness and sometimes death. A risk factor is heat exposure or athletic overexertion, since dopamine blockade impairs hypothalamic heat regulation. Often intensive care hospitalization is required. Dantrolene is often an effective treatment. It is important to diagnose early, and serum CPK is useful.

The metabolic side effects of the second generation of antipsychotics were unforeseen, despite their presence in clozapine which inspired the synthesis and creation of the second generation as a whole [31]. Clozapine itself has powerful weight increasing effects and diabetogenic effects. Its effect to cause agranulocytosis in 1–8% of those taking it has led to a strict regime in which patients are evaluated weekly for white blood count for the first 6 months and monthly thereafter for the rest of their clozapine therapy. Clozapine is therefore not a first line treatment but its use in resistant psychosis is an essential component of the psychopharmacological armamentarium. Importantly, clozapine seems to be as impressively effective in resistant severe bipolar disorder as it is in resistant schizophrenia-like psychoses. It has also been reported to be useful in resistant psychotic depression and in resistant severe OCD. Its difficult side effect profile combined with its unique therapeutic usefulness in highly mentally ill patients presents a psychiatrist with the kind of dilemma often faced by the oncologist or cardiologist and makes his medical training in difficult clinical situations meaningful and essential.

The mechanism by which clozapine and many second generation antipsychotics developed subsequently such as olanzapine and quetiapine to increase weight and decrease glucose tolerance is not understood. Many hypotheses exist. The weight gain is mediated by increased caloric intake and could be due to effects on insulin or glycogen receptors both inside or outside the brain. However, there seem to be effects on glucose tolerance over and above the weight gain and also effects to increase atherogenic blood lipids above and beyond the effects to increase weight gain and appetite. Using appropriate animal models and hypotheses about specific serotonin receptor mediation of appetite and glucose control, aripiprazole and lurasidone have been developed and are probably weight neutral. However, the full extent of their clinical usefulness is not yet apparent and given the faddish nature of prescribing in this area (Fig. 6.4) perhaps discretion is the better part of valor for the prescribing physician.

The use of modern imaging techniques has revealed worrisome effects of long-term antipsychotic exposure to cause reduced brain volume, and this has been replicated in animal studies [32]. The effect is not limited to first generation medications, and has been well shown with olanzapine. It is not clear yet if some antipsychotics are devoid of this effect, or if it is indeed harmful. In normal adolescence brain volume of some brain areas declines as part of the pruning of unwanted synapses. Moreover, the brain is mostly water, and changes in water distribution could explain some of the effects of antipsychotics and may be reversible or even benign. Clinical treatment must be a weighing of benefit vs harm in a situation of inadequate current data, but further studies on this issue are underway.

Given the proliferation of medicines to treat COVID in the current era, it is important that physicians check for possible interactions with COVID therapy when prescribing antipsychotic drugs [33].

Clinical Vignettes

1. Robert was a 46 year old single man who presented to the Psychiatry Outpatient Clinic with anxiety. He had dropped out of college at age 22 and supported himself with odd jobs and never had a serious relationship. At age 27 he had a psychosis treated with antipsychotics that he discontinued shortly after his brief hospitalization. Because of the deterioration in function from his late adolescent level, he was diagnosed as schizophrenic on hospital discharge. At age 46 he was evaluated and the resident psychiatrist wished to send him to the long-acting neuroleptic follow-up clinic. The supervising psychiatrist suggested a new psy-

chosocial crisis intervention program with 10 meetings and a low dose benzodiazepine. He did well.

2. Gerry was a 37 year old homeless man with chronic psychosis who avoided
medical contact because he was usually sent for a long-acting neuroleptic injection whenever he appeared at an emergency room, from which he had no known
benefit but many memorable side effects. A new program offered him housing
and a non-judgmental attitude toward his psychotic ideas. His nutrition, dental
hygiene and quality of life improved noticeably.

3. Gill is a 40 year old woman with recurrent psychotic attacks since age 18, poor
function interpersonally, dependence on her family and some paranoid aggression toward her neighbours. She is maintained on antipsychotic treatment, without which she is much worse and uncontrollable at home as she has demonstrated
on numerous occasions when she stops her medication. She is never psychosis
free and on periodic follow-up her medication dose was usually increased with
the stated aim in her chart of "recovery". In practice, the dose increase usually
led to cessation of her menses, galactorrhea, akathisia and rehospitalization. A
doctor's seminar on goals and dose optimization led to a new regime where psychosis amelioration was balanced with side effects as a chart goal in this particular clinic. She remained at home with behavior acceptable to her family on low
dose neuroleptic (although always a bit psychotic) for a much longer period of
time than in the past.

References

1. Ban A. Psychopharmacology. Williams & Wilkins Company; 1969.
2. Klein D, Davis J. Diagnosis and drug treatment of psychiatric disorders. 1st ed. Williams & Wilkins; 1969.
3. Shorter E. The rise and fall of the age of psychopharmacology. Oxford University Press; 2021.
4. Li M, Fletcher PJ, Kapur S. Time course of the antipsychotic effect and the underlying behavioral mechanisms. Neuropsychopharmacology. 2007;32(2):263–72.
5. Goff DC. The pharmacologic treatment of Schizophrenia-2021. JAMA. 2021;325(2):175–6.
6. Siafis S, Davis JM, Leucht S. Antipsychotic drugs: from 'major tranquilizers' to neuroscience-based-nomenclature. Psychol Med. 2021;51(3):522–4.
7. McCutcheon RA, Pillinger T, Mizuno Y, Montgomery A, Pandian H, Vano L, et al. The efficacy and heterogeneity of antipsychotic response in schizophrenia: a meta-analysis. Mol Psychiatry. 2021;26(4):1310–20.
8. Lerner Y, Lwow E, Levitin A, Belmaker RH. Acute high-dose parenteral haloperidol treatment of psychosis. Am J Psychiatry. 1979;136(8):1061–4.
9. Carpenter WT. What is missing in treatment guidelines? Schizophr Bull. 2021;47(2):269–70.
10. Moncrieff J. Antipsychotic maintenance treatment: time to rethink? PLoS Med. 2015;12(8):e1001861.
11. Siafis S, Deste G, Ceraso A, Mussoni C, Vita A, Hasanagic S, et al. Antipsychotic drugs v. barbiturates or benzodiazepines used as active placebos for schizophrenia: a systematic review and meta-analysis. Psychol Med. 2020;50(15):2622–33.

12. Huhn M, Nikolakopoulou A, Schneider-Thoma J, Krause M, Samara M, Peter N, et al. Comparative efficacy and tolerability of 32 oral antipsychotics for the acute treatment of adults with multi-episode schizophrenia: a systematic review and network meta-analysis. Lancet. 2019;394(10202):939–51.
13. Creese I, Burt DR, Snyder SH. Dopamine receptor binding predicts clinical and pharmacological potencies of antischizophrenic drugs. Science. 1976;192(4238):481–3.
14. Wyatt RJ, Murphy DL, Belmaker R, Cohen S, Donnelly CH, Pollin W. Reduced monoamine oxidase activity in platelets: a possible genetic marker for vulnerability to schizophrenia. Science. 1973;179(4076):916–8.
15. Belmaker RH. The lessons of platelet monoamine oxidase. Psychol Med. 1984;14(2):249–53.
16. Kane J, Honigfeld G, Singer J, Meltzer H. Clozapine for the treatment-resistant schizophrenic. A double-blind comparison with chlorpromazine. Arch Gen Psychiatry. 1988;45(9):789–96.
17. Ebstein RP, Novick O, Umansky R, Priel B, Osher Y, Blaine D, et al. Dopamine D4 receptor (D4DR) exon III polymorphism associated with the human personality trait of novelty seeking. Nat Genet. 1996;12(1):78–80.
18. Lieberman JA, Stroup TS, McEvoy JP, Swartz MS, Rosenheck RA, Perkins DO, et al. Effectiveness of antipsychotic drugs in patients with chronic schizophrenia. N Engl J Med. 2005;353(12):1209–23.
19. Sokoloff P, Le Foll B. The dopamine D3 receptor, a quarter century later. Eur J Neurosci. 2017;45(1):2–19.
20. Luykx JJ, Tiihonen J. Antipsychotic discontinuation: mind the patient and the real-world evidence. Lancet Psychiatry. 2021;8(7):555–7.
21. Kapur S. Psychosis as a state of aberrant salience: a framework linking biology, phenomenology, and pharmacology in schizophrenia. Am J Psychiatry. 2003;160(1):13–23.
22. Moncrieff J, Cohen D, Mason JP. The subjective experience of taking antipsychotic medication: a content analysis of internet data. Acta Psychiatr Scand. 2009;120(2):102–11.
23. Begemann MJH, Thompson IA, Veling W, Gangadin SS, Geraets CNW, van't Hag E, et al. To continue or not to continue? Antipsychotic medication maintenance versus dose-reduction/discontinuation in first episode psychosis: HAMLETT, a pragmatic multicenter single-blind randomized controlled trial. Trials. 2020;21(1):147.
24. Chouinard G, Samaha AN, Chouinard VA, Peretti CS, Kanahara N, Takase M, et al. Antipsychotic-induced dopamine Supersensitivity psychosis: pharmacology, criteria, and therapy. Psychother Psychosom. 2017;86(4):189–219.
25. Whitaker R. Mad in America: bad science, bad medicine, and the enduring mistreatment of the mentally ill. Basic Books; 2019.
26. Kishi T, Ikuta T, Matsuda Y, Sakuma K, Okuya M, Mishima K, et al. Mood stabilizers and/or antipsychotics for bipolar disorder in the maintenance phase: a systematic review and network meta-analysis of randomized controlled trials. Mol Psychiatry. 2021;8(26):4146–57.
27. Kishi T, Matsuda Y, Sakuma K, Okuya M, Mishima K, Iwata N. Recurrence rates in stable bipolar disorder patients after drug discontinuation v. drug maintenance: a systematic review and meta-analysis. Psychol Med. 2021;51(15):2721–9.
28. Bushnell GA, Crystal S, Olfson M. Trends in antipsychotic medication use in young privately insured children. J Am Acad Child Adolesc Psychiatry. 2021;60(7):877–86.
29. Arango C, Díaz-Caneja CM, McGorry PD, Rapoport J, Sommer IE, Vorstman JA, et al. Preventive strategies for mental health. Lancet Psychiatry. 2018;5(7):591–604.
30. Belmaker RH, Wald D. Haloperidol in normals. Br J Psychiatry. 1977;131:222–3.
31. Sneller MH, de Boer N, Everaars S, Schuurmans M, Guloksuz S, Cahn W, et al. Clinical, biochemical and genetic variables associated with metabolic syndrome in patients with schizophrenia Spectrum disorders using second-generation antipsychotics: a systematic review. Front Psych. 2021;12:625935.
32. Voineskos AN, Mulsant BH, Dickie EW, Neufeld NH, Rothschild AJ, Whyte EM, et al. Effects of antipsychotic medication on brain structure in patients with major depressive disorder and

psychotic features: neuroimaging findings in the context of a randomized placebo-controlled clinical trial. JAMA Psychiatry. 2020;77(7):674–83.

33. Plasencia-García BO, Rico-Rangel MI, Rodríguez-Menéndez G, Rubio-García A, Torelló-Iserte J, Crespo-Facorro B. Drug-drug interactions between COVID-19 treatments and antidepressants, mood stabilizers/anticonvulsants, and benzodiazepines: integrated evidence from 3 databases. Pharmacopsychiatry. 2021;238:329–40.

Chapter 7
Antianxiety Medications: Are They Addictive or Are They Mankind's Precious Heritage?

Historical Background

Anxiety is a feeling of impending doom, a perception of heart palpitations, knee shaking, dry mouth and intense discomfort that can come in waves or exist as a background feeling continuously in an individual in the absence of an external realistic threat or reason for fear. Anxiety is intensely uncomfortable and interferes with pleasurable activities or the concentration necessary for consistent work productivity. In its many forms, it is very prevalent in the population and has been described since ancient times. It is possible, but not proven, that the electronic stimuli of modern life have increased the prevalence or intensity of anxiety but this is difficult to prove. There is no evidence that human beings once existed in a "natural" Rousseau-like state of harmony with nature without anxiety. Mankind's history shows that substances such as alcohol were used many thousands of years ago, perhaps often to reduce unwanted anxieties. Anxiety can be relieved in many individuals by intentional muscle relaxation, by rhythmic breathing exercises, by distractions including music or exercise, by pleasurable activities including eating and sexual activities and by concentrating on pleasurable or distracting thoughts. Some people are not able to control their own anxiety and the resulting discomfort and disability leads them to seek help. Several forms of non-pharmacological treatment are proven to be helpful in anxiety states and will not be discussed in this chapter. However, pharmacological treatments of anxiety must always be evaluated, both in research and in the clinical situation, compared with alternative non-pharmacological treatments. The individual's complaint of anxiety should be seen in the context of the millennia-old human quest for relief of anxiety and the sufferer of anxiety should not feel guilty of moral imperfection or made to feel that if he were living a more moral or healthy life style that he would of necessity not be having his complaint. Philosophical angst or political panic are not treatable by pharmacological methods.

© The Author(s), under exclusive license to Springer Nature Switzerland AG 2023
R. H. Belmaker, P. Lichtenberg, *Psychopharmacology Reconsidered*,
https://doi.org/10.1007/978-3-031-40371-2_7

The Nonlinear Nature of Anti-anxiety Treatment

A basic approach to the treatment of anxiety uses the famous Yerkes-Dodson curve presented in Fig. 7.1. Figure 7.1 is called the inverted u-curve and is easily demonstrated in laboratory rodents where function can be measured by the amount of time necessary to find their way through a moderately complex maze. Rats will stroll through such a maze and if they find food at the other end will move more quickly the next time they are put at the beginning of the maze and will remember many of the turns it took to successfully find the food. If given a small slightly painful electric shock before starting the maze, the rat is mentally and autonomically more aroused and will actually complete the maze more quickly. However, if increasingly stronger electrical shocks are given an optimal point is reached as shown in the peak of Fig. 7.1 and increasing levels of pain lead to increases in the amount of time that the rat takes to perform in the maze, indicating decreasing function with increasing anxiety. In human terms this curve describes the old adage that a little bit of anxiety is good for function but too much is harmful to human function. These findings have been replicated in most human functions. In the work place, the apathetic worker with little motivation is a problem in business administration textbooks. The worker with too much anxiety becomes a problem worthy of medical attention [1].

Alcohol as a Proto Anti-anxiety Drug

Alcohol use as a drug has many drawbacks: Its half life is short, it has long term negative physical effects including those on the liver. It affects different parts of the brain at different rates such that in many individuals it reduces cortical function first and leads to disinhibition before reducing anxiety. It has a narrow therapeutic index and too much can cause coma or death. The first medical anxiolytic was phenobarbital which was synthesized early in the nineteenth century. It became very widely used and chemical permutations of the molecule such as secobarbital, thiopentone

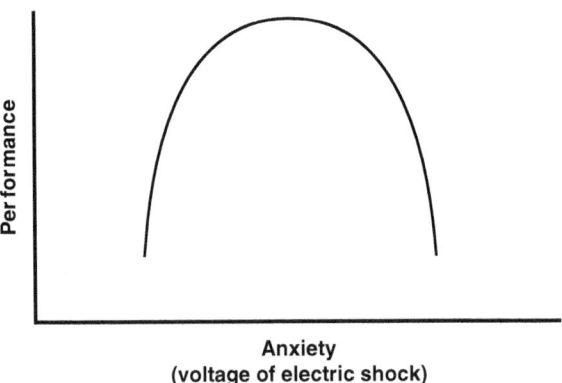

Fig. 7.1 The inverted U relationship between anxiety and function

and others were easily derived from the parent compound. These compounds were found to have three related effects which came to be conceptualized separately: (a) At higher doses, these compounds could induce sleep in patients complaining of insomnia [2]. (b) At lower doses these compounds relieved daytime anxiety in patients complaining of anxiety. (c) At very high doses these compounds could be used to sedate patients in acute psychotic states or violent states of varied origin [3]. These three properties were called respectively (a) the hypnotic effect, (b) the anxiolytic effect and (c) the sedative effect. An additional effect of these compounds in neurology was a marked effect to treat and prevent epileptic seizures, an effect still used today.

The barbiturate sedative/anxiolytic/hypnotic compounds were rapidly found to have three major medical problems. (1) In overdoses they all depressed the respiratory center in the brain and led to death by respiratory depression. (2) In daily dosage they all led to the development of tolerance within months or years and doses needed to be increased to achieve the same effect. (3) If doses were increased to maintain an effect over years, rapid cessation of treatment with these compounds led to a withdrawal syndrome of anxiety, insomnia, agitation and epileptic seizures. The withdrawal syndrome was the mirror image of the treatment effect of these compounds. Fear of a withdrawal effect combined with the need of increasing doses to achieve the anxiolytic and hypnotic effects of these compounds led to a pharmacological and behavioral syndrome called addiction. The field of the psychopharmacology of anxiety since barbiturates has been characterized by the search for a non-addicting anxiolytic and hypnotic which did not depress respiratory function on overdose and which could not be then considered a risk for suicide.

Non-barbiturate Sedatives

The pharmaceutical industry worked hard to develop such compounds in the 1940s and 1950s. One of the first to be developed was meprobamate which was highly promoted in the United States for "the bored housewife". This sexist and arrogant approach to the commercialization of antianxiety agents has been well documented in historical accounts. Meprobamate was at first advertised as non-addicting and not fatal in overdose; these claims were disproven within a year. Meanwhile, a new non barbiturate sedative and anxiolytic methaqualone was synthesized and promoted. This process was repeated almost yearly for over a decade. These compounds are almost unknown in clinical practice today.

Benzodiazepines

The first benzodiazepine was discovered in Roche Pharmaceuticals in Nutley New Jersey, chlordiazepoxide (Librium) It was so successful commercially that it became a cultural icon. The benzodiazepines were synthesized in a semi-rational manner because their discoverer Sternbach was scanning compounds for effects on mouse behavior. Librium and its congener diazepam (valium) have clear anxiolytic effects on daytime anxiety. They were also effective sleep promoting compounds if given at night, perhaps at slightly higher doses. They were perhaps not as good as barbiturates for acute sedation in psychosis or violence but this was a rarer indication as antipsychotics became available (see Chap. 6). Over their early use it became clear that it was indeed much more difficult to achieve respiratory depression in overdose with benzodiazepines than it had been with barbiturates and that successful suicide was and has been rare. However, there was no absolute distinction since many patients overdosed with combinations of benzodiazepines and alcohol or benzodiazepines and other drugs such that benzodiazepines should not be considered to be completely without suicide risk. Tolerance and addiction are less common with benzodiazepines than with barbiturates; however, both tolerance and addiction certainly occur and must be taken into account by the treating clinician. A particular problem is alprazolam, a very short acting benzodiazepine that can give anxiety relief within minutes but is metabolized within a couple of hours. The rapid effect and rapid return of symptoms can generate a conditioned response that can progress to addiction. Longer acting agents such as clonazepam are more likely to lose pharmacological effect after the acute anxiety attack has finished its natural course and so avoid the immediate rebound of symptoms, and should be preferred [4].

After the discovery of diazepam numerous pharmaceutical companies synthesized similar compounds in the benzodiazepine family. The goal of many of these syntheses was to develop compounds with specific half-lives appropriate for giving a night's sleep, for immediate absorption and rapid excretion or metabolism, or for longer term stable dose treatment. Some of the compounds were more lipophilic and others less lipophilic. For instance, a short half-life compound which is lipophilic will gradually be absorbed in the body's fatty tissues and if given repeatedly will be slowly released in between doses or upon cessation of dosing. Therefore, it will under certain circumstances behave like a long half life compound. Thus, complex pharmacokinetic discussions have been written about benzodiazepines. The alpha half life and the beta half-life and their clinical implications are discussed below.

As of today, several dozen different benzodiazepines have been brought to market and there were periods of intense marketing competition for the treatment of anxiety and insomnia. Almost all of these compounds are now off patent and the remaining most often used benzodiazepines are clonazepam, lorazepam and diazepam [5].

The Mechanism of Action of Benzodiazepines

The elucidation of the mechanism of action of benzodiazepines was a world-wide scientifically heroic and exciting story, which has been documented in several accounts. While commercial competition between companies played a role, academic basic research took the lead. The benzodiazepine compound acts at a receptor in brain cells that is intriguingly and surprisingly the receptor for no proven endogenous ligand. It is part of the huge receptor for the natural neurotransmitter GABA as shown in Fig. 7.2. This receptor has numerous components that could only be compared to a Rube Goldberg apparatus. Its main function is to bind GABA that is released from the presynaptic neuron and, when GABA is bound, to open an adjacent chloride channel allowing chloride ions to flow into the post synaptic cell and depolarize the neuron. A benzodiazepine drug does not directly affect the binding of GABA in an inhibitory or competitive way. The benzodiazepine receptor is allosterically linked to the GABA binding site such that when the benzodiazepine receptor is occupied, GABA binds more strongly and is more effective in opening the chloride channel. Thus, benzodiazepines can only function in the presence of GABA. The GABA neurotransmission system has been with us since the development of early multicellular animals with rudimentary nervous systems.

The effect of GABA after opening the chloride channel is to depress firing of the post synaptic cell. It is thus an inhibitory neurotransmitter as discussed in Chap. 2. By facilitating GABA neurotransmission, benzodiazepine binding to the benzodiazepine receptor enhances this inhibition. Since 40% of brain neurons function using GABA, numerous brain functions can be inhibited by benzodiazepine's binding to

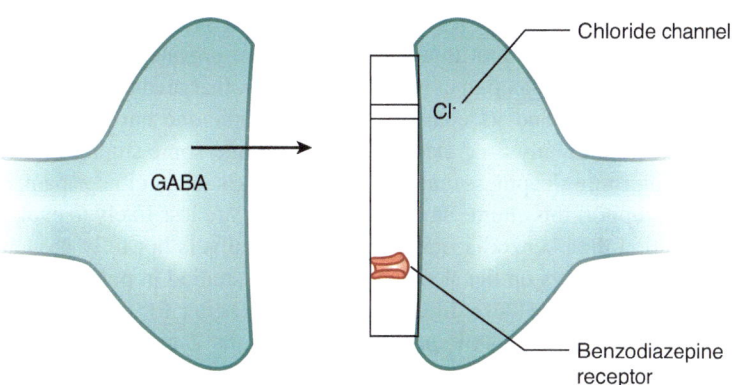

Fig. 7.2 The benzodiazepine receptor is spart of a larger complex GABA receptor

its receptor on the GABA complex. For instance, it is not surprising that these substances that inhibit firing of 40% of brain neurons induce sleep. It is also not surprising that they depress the frequency by which an epileptic focus in the brain expands its activity to cause a full blown seizure. It could also be understood why these medicines reduce daytime anxiety and also clear why they reduce reaction time and impair motor vehicle driving ability while doing so (see below in side effects).

Overdoses of benzodiazepines causing prolonged sleep or coma can be reversed using benzodiazepine receptor antagonists. The most common benzodiazepine antagonist is flumazenil [6] which should be available in every hospital emergency room. A more complex concept at the benzodiazepine receptor is that of the inverse agonist. These compounds affect the benzodiazepine receptor to decrease GABA binding and cause intense and intolerable anxiety in normal human volunteers. They provide an example of where a complex human emotion can be induced by action at a specific molecularly well-defined site.

The Z Drugs

While benzodiazepines were effective in the three components originally conceptualized for the barbiturates, that is anxiolytic, hypnotic and sedative effects, a significant number of patients complained of daytime fogginess in the morning after use of a benzodiazepine as a sleeping medication. Attempts to develop different half-life compounds were not entirely successful as a three-hour half-life was too short and an eight hour half-life left too much medicine in the blood stream in the morning. Synthetic chemists, and in particular Solomon Langer, based in the French chemical company SyntheLabo, developed a compound called zolpidem that they claimed bound more effectively to one variant of a subunit of the benzodiazepine receptor. They claimed that this variant was more prevalent in those areas of the brain responsible for sleep than in the widespread other areas of GABA neurotransmission. Zolpidem and its congeners such as zopiclone have become extremely popular and commonly used sleep medicines. While some studies report that they have less morning sleepiness than a benzodiazepine such as clonazepam, other studies do not support this claim. Studies find that they are less likely to evoke tolerance and addiction than benzodiazepines for sleep but this is not fully proven. They do have different effects on the sleep architecture as studied in polysomnography, but it is not clear whether these differences are really better for the patient or not. The initial biochemical claims of specific binding of the Z compounds to a particular functional group of GABA associated benzodiazepine receptors is not completely proven or disproven today [7, 8].

Comparison of Treatment of Anxiety with Benzodiazepines Vs. Antidepressants

In Chap. 5 we discussed the use of monoamine reuptake blockers in the treatment of anxiety and their effectiveness in panic disorder. An increasing number of studies suggest that these compounds have effectiveness in some cases of general anxiety disorder as well. The antidepressant drugs were heavily promoted by commercial interests at the time that benzodiazepines began to go off patent. Many academic researchers have written that the treatment of anxiety with antidepressant compounds is more specific and less likely to cause tolerance and addiction than treatment with benzodiazepines. Our clinical experience and reading of the current literature does not support this view. Benzodiazepines still play a major role in the armamentarium of the psychiatrist as well as the family physician dealing with insomnia or daytime anxiety. The advantages of benzodiazepine treatment of anxiety are as follows:

1. A patient can take a benzodiazepine as needed when he feels anxiety is coming on and is extremely likely to achieve prevention or rapid relief. In periods when he is feeling better, he need not take the treatment. Antidepressant treatment of anxiety disorder requires daily preventive treatment and often patients continue in periods when they could have stopped taking any medication at all [9, 10].
2. Many patients with anxiety feel a need to be in control of their anxiety and benzodiazepines can be administered on an as-needed basis where the patient decides himself when he needs a tablet. Often situations evolve to where patient finds that a diazepam tablet is 100% effective in the patient's pocket; he never needs to take it if he knows that he has one when and if he might need one. This situation does not develop with antidepressant treatment.
3. It is increasingly claimed that antidepressant treatment has general effects on the personality that may be as worrisome as the drowsiness that can be caused by benzodiazepine treatment.
4. Benzodiazepines interact only with GABA and GABA is present only in the central nervous system. They have no effects on the heart, no effects to cause GI constipation, no danger of causing urinary retention, blurred vison, nausea or sexual side effects [11].
5. The dose of benzodiazepines is extremely flexible and can be changed by the patient with telephone conversation with the physician within almost an order of magnitude, increasing in periods of stress and anxiety and decreasing whenever possible. By contrast, antidepressants are given within a narrow dose range and the dose cannot be adjusted depending on the daily situation but only at frequencies of approximately 3 weeks or a month.

6. Many patients who start antidepressant treatment for anxiety have an exacerbation of their anxiety at the beginning of the antidepressant treatment and need benzodiazepines for several weeks at least. They are thereby exposed to benzodiazepines and find them effective and often continue them along with their antidepressant medication. If the physician starts off with monotherapy of benzodiazepines, the long term effect might be less drug usage rather than using both drugs if antidepressants are started first.

In the clinic our practice is to use antidepressants only for defined panic disorder and otherwise to manage anxiety with benzodiazepines for the daytime and Z drugs for nighttime insomnia. It is critical for the family physician or the psychiatrist to be available by telephone, whatsapp, or email and to encourage the patient to take medicine when needed and to reduce the drug or stop it when anxiety is controlled or ends its natural course of illness. Many patients are afraid of becoming addicted to benzodiazepines but if they do not take the medicine their anxiety will impair their quality of life and negatively affect their prognosis. On the other hand, patients who do take benzodiazepines often find it such a positive relief of an unpleasant anxious emotion that they find it difficult to give up even when they don't need it: the physician must be there to encourage weaning and withdrawal at the proper time [12]. Needless to say, non-pharmacological approaches such as CBT and mindfulness also have demonstrated effectiveness in anxiety disorders and should be used whenever possible instead of or in conjunction with benzodiazepines. However, the physician who insists on recommending CBT first for patients who do not have the verbal ability for it, do not have the patience for it or who do not have access to CBT in their neighborhood is not achieving a morally higher level by deriding the psychopharmacological approach.

Side Effects

Benzodiazepines have few side effects outside of the brain or spinal cord since GABA exists only in the central nervous system. Their acute side effects in the central nervous system relate to drowsiness or reduced alertness and the subsequent impairment of ability to operate motor vehicles or stay awake [13]. For treatment of insomnia, drowsiness the following morning should be handled by finding the compound with the appropriate half-life for the specific patient. It should be noted that half-lives determined in the scientific literature are averages for normal volunteers and that inter-individual differences for the same drug are often larger than inter-drug differences for the average half-life quoted in the literature. Therefore, individualization of treatment by the physician is the key. Since anxiety waxes and wanes, as does insomnia for most patients, the physician must remain in contact with the patient by phone, whatsapp or by email to adjust doses upward if necessary and downward whenever possible. The office prescription with indefinite renewals is a disaster in this area as in many other areas of psychiatry and medicine.

Large data bases of hundreds and thousands or even millions of patients often find correlations between benzodiazepine use and all cause mortality in groups of patients, especially the elderly [14–16]. Sometimes connections are reported that are biologically implausible because they do not occur in any animal model with any conceivable mechanism. It should be remembered that association is not causality and it is possible that patients with serious medical illnesses, including difficult to define ones such as frailty, may turn more often to their physicians for the treatment of anxiety and thereby receive benzodiazepines. The association between benzodiazepine use and falls and fractures in the elderly is more substantiated and more logical than some of the other reported associations. The risk of falls should be emphasized to the patient and family and the access from bed to bathroom be free of stumbling blocks or banana peels. The risk of such falls must be judged against the benefit of anti-anxiety medication or sleep medication for the individual patient by his individual physician. Toleration or ignoring intense discomforting anxiety is not an option for most patients. Obviously, casual or thoughtless prescription of benzodiazepines as a way of escorting a troublesome patient out of the office is not good medicine either.

The specific advice on use of benzodiazepines in patients with a driver's license depends on the specific compound half-life, responsibility in its use by a specific patient and the laws and legal precedents regulating medical liability in the physician's specific jurisdiction. It should be clear that all of medicine is a risk/benefit undertaking with no free lunch.

Sleep Architecture

Sleep is a critical psychophysiological function and human mood and social interactions are greatly impaired if sleep is chronically or frequently impaired. Sleep is a multiphase process and includes dream, nondream, deep sleep and light sleep components. Surprisingly, most antidepressants have strong effects to suppress REM or dream sleep whereas benzodiazepines or Z drugs have few or no effects on sleep architecture even as they decrease time to falling asleep and increase total sleep duration. Some short acting benzodiazepines and perhaps most often reported, zolpidem, are associated with night-time events involving waking up with motor activity but amnesia the next day for the event. It is hard to give this phenomenon a name without implying a judgement on causality or mechanism. Most likely the phenomenon is no different than that seen when waking up a person in a deep sleep without pharmacological exposure. There can be confusion and motor activity with later amnesia. Sometimes the phenomenon after use of sleep medication can have medicolegal implications [17].

Benzodiazepine Use in Psychiatry-Related Medicine

Ultrashort acting benzodiazepines can be used in anesthesia before ECT. Effects on seizure threshold are managed clinically. Chlordiazepoxide, the first benzodiazepine, is still a preferred treatment in withdrawal from alcohol addiction and in treatment of delirium tremens [18], the alcohol withdrawal psychosis. Very high doses are necessary and should not be lowered quickly. The cross-tolerance of alcohol and benzodiazepines in addiction and withdrawal returns the reader to the beginning of this chapter, where the millennium-old use of alcohol by humanity must be understood in forming a philosophy of benzodiazepine use in anxiety. Neurologists routinely use some benzodiazepines, such as clonazepam, as well as the barbiturate phenobarbital, in the long term treatment of epilepsy. Muscle relaxant effects of benzodiazepines, mediated by spinal neurons and not direct effects on muscle, are useful in treatment of tension headache and peripheral muscle spasms but should be used sparingly in chronic back pain where tolerance and addiction are more often reported. Intravenous diazepam is effective in acute neuroleptic-induced dystonia and is more rapid than intramuscular anticholinergic treatment; however, an ambu bag should be available if apnea occurs [19].

Pregnancy and Lactation

Benzodiazepines and Z compounds do not have specific known teratogenicity. High dose use close to delivery may cause sleepy baby syndrome. Many benzodiazepines pass into breast milk, and the specific compound considered for treatment of the lactating mother should be checked in the literature before use. As with all psychoactive compounds, long term effects on the developing brain in utero or in the breastfeeding period are not known and use should be minimized to those mothers in absolute need [20, 21].

Clinical Vignettes

1. Robert is a 42 year old man who arrived at psychiatry clinic after 2 months of almost total sleeplessness since being fired from his job of 15 years. He had failed three job interviews because of his exhausted, somewhat disheveled and stuttering appearance. After 10 days on zopiclone 7.5 mg before bed he had gotten a full night's sleep nightly and organized his appearances so that he was successfully rehired and within a week had stopped his sleep medicine.
2. Jane was a 32 year old junior university lecturer with incapacitating anxiety before public speaking. She tried to avoid frontal lecture assignments but it was becoming impossible. She was prescribed 0.5 mg clonazepam to try first on

weekends before mock lectures to her family and the dose adjusted to avoid sleepiness. She successfully got through her first 2 months of the semester taking 0.25 mg clonazepam a half hour before each lecture and then went down to 0.125 mg clonazepam. Subsequently she always kept a pill in her purse but rarely had to use it.

3. Joshua was a 67 year old man who lost his closest brother in an auto accident and was so unable to sleep that he could not receive visitors who appeared at his home for the shiva (mourning week). There was no way that he was using his religious traditions or rituals to work through his loss because of his insomnia and anxiety. He was sedated with 0.5 mg clonazepam three times daily and 1 mg before bed. The dose was tapered over 2 weeks before the ritual 1 month visit to the cemetery with family and he was able to achieve a normal bereavement process.

4. Clara was a 37 year old woman who had a single car auto accident with some broken ribs and arrived in the emergency room in a daze completely unable to concentrate on simple tasks. She was hyperventilating and had a gross tremor. Her neurological examination and tests were normal. She was sedated with 20 mg of diazepam orally and recovered her composure well. She was sent home on 20 mg diazepam daily, unfortunately without phone follow-up and dose reduction. She arrived in the emergency room 10 days later in a light coma. The drug had accumulated in her fatty tissues and took almost 3 days before she was again functional because of the long beta half life of the drug.

5. Joseph was a 22 year old man who had been married 2 years before and had not shown interest in starting a job to support a family. He complained of headaches and anxiety because of his wife's shouting and had not responded to maximal doses of benzodiazepines and headache pain medicine that the family physician felt able to provide. On psychiatric consultation he was advised to look actively for work, stop benzodiazepines and was offered a joint meeting with the wife to make a family financial plan including employment for Joseph.

References

1. Craske MG, Stein MB. Anxiety. Lancet. 2016;388(10063):3048–59.
2. Morin CM, Benca R. Chronic insomnia. Lancet. 2012;379(9821):1129–41.
3. Lerner Y, Lwow E, Levitin A, Belmaker RH. Acute high-dose parenteral haloperidol treatment of psychosis. Am J Psychiatry. 1979;136(8):1061–4.
4. Soyka M. Treatment of benzodiazepine dependence. N Engl J Med. 2017;376(12):1147–57.
5. Kurko TA, Saastamoinen LK, Tähkäpää S, Tuulio-Henriksson A, Taiminen T, Tiihonen J, et al. Long-term use of benzodiazepines: definitions, prevalence and usage patterns - a systematic review of register-based studies. Eur Psychiatry. 2015;30(8):1037–47.
6. Penninga EI, Graudal N, Ladekarl MB, Jürgens G. Adverse events associated with flumazenil treatment for the management of suspected benzodiazepine intoxication - a systematic review with meta-analyses of randomised trials. Basic Clin Pharmacol Toxicol. 2016;118(1):37–44.
7. Murrough JW, Yaqubi S, Sayed S, Charney DS. Emerging drugs for the treatment of anxiety. Expert Opin Emerg Drugs. 2015;20(3):393–406.

8. Wright BT, Gluszek CF, Heldt SA. The effects of repeated zolpidem treatment on tolerance, withdrawal-like symptoms, and GABAA receptor mRNAs profile expression in mice: comparison with diazepam. Psychopharmacology. 2014;231(15):2967–79.
9. Ogawa Y, Takeshima N, Hayasaka Y, Tajika A, Watanabe N, Streiner D, et al. Antidepressants plus benzodiazepines for adults with major depression. Cochrane Database Syst Rev. 2019;6(6):Cd001026.
10. Slee A, Nazareth I, Bondaronek P, Liu Y, Cheng Z, Freemantle N. Pharmacological treatments for generalised anxiety disorder: a systematic review and network meta-analysis. Lancet. 2019;393(10173):768–77.
11. Stein MB, Sareen J. Clinical practice. Generalized anxiety disorder. N Engl J Med. 2015;373(21):2059–68.
12. Olfson M, King M, Schoenbaum M. Benzodiazepine use in the United States. JAMA Psychiatry. 2015;72(2):136–42.
13. Dubois S, Bédard M, Weaver B. The impact of benzodiazepines on safe driving. Traffic Inj Prev. 2008;9(5):404–13.
14. Verdoux H, Lagnaoui R, Begaud B. Is benzodiazepine use a risk factor for cognitive decline and dementia? A literature review of epidemiological studies. Psychol Med. 2005;35(3):307–15.
15. Saarelainen L, Taipale H, Koponen M, Tanskanen A, Tolppanen AM, Tiihonen J, et al. The incidence of benzodiazepine and related drug use in persons with and without Alzheimer's disease. J Alzheimers Dis. 2016;49(3):809–18.
16. Billioti de Gage S, Moride Y, Ducruet T, Kurth T, Verdoux H, Tournier M, et al. Benzodiazepine use and risk of Alzheimer's disease: case-control study. BMJ. 2014;349:g5205.
17. Louzada LL, Machado FV, Quintas JL, Ribeiro GA, Silva MV, Mendonça-Silva DL, et al. The efficacy and safety of zolpidem and zopiclone to treat insomnia in Alzheimer's disease: a randomized, triple-blind, placebo-controlled trial. Neuropsychopharmacology. 2022;47(2):570–9.
18. Schuckit MA. Recognition and management of withdrawal delirium (delirium tremens). N Engl J Med. 2014;371(22):2109–13.
19. Gagrat D, Hamilton J, Belmaker RH. Intravenous diazepam in the treatment of neuroleptic-induced acute dystonia and akathisia. Am J Psychiatry. 1978;135(10):1232–3.
20. Bais B, Molenaar NM, Bijma HH, Hoogendijk WJG, Mulder CL, Luik AI, et al. Prevalence of benzodiazepines and benzodiazepine-related drugs exposure before, during and after pregnancy: a systematic review and meta-analysis. J Affect Disord. 2020;269:18–27.
21. Bellantuono C, Tofani S, Di Sciascio G, Santone G. Benzodiazepine exposure in pregnancy and risk of major malformations: a critical overview. Gen Hosp Psychiatry. 2013;35(1):3–8.

Chapter 8
Mood Stabilizers: Off the Gold Standard?

The discovery of lithium's therapeutic potential preceded the introduction in the 1950s of the first neuroleptics, but was overshadowed by the latter, which were easier to use, did not require blood tests, had a wider therapeutic index, and could be patented. Nevertheless, lithium as a simple ion, extracted from the earth as an ore, neither synthesized in a factory nor metabolized by the body, yet providing therapeutic benefits, quickly became a source of fascination in psychopharmacology. The fact that lithium might interact with the electrophysiology of neurotransmission along with sodium and potassium intrigued the scientific community, which hoped that lithium might hold the key to unlocking the mysteries of the basic biochemistry and neurophysiology of manic depressive illness, as bipolar disorder was then called.

Pharmacokinetics and Pharmacodynamics

Lithium is completely absorbed via the gastrointestinal tract into the blood stream within hours ingestion, whether taken before, after, or without meals. Excretion is via the kidney into the urine. Lithium is water soluble and lipophobic. To have its effect, lithium must cross the lipid membrane of neurons in the brain into the cytoplasm in brain cells, which it apparently does without interacting directly with extracellular receptors. Since the various lithium salts—chloride, carbonate, sulfate, orotate, and citrate—all dissolve in the aqueous medium of the stomach, lithium is absorbed equally well in all these forms. Converting a simple salt into a "slow-release" preparation is technically difficult, so that commercial claims of "slow release" lithium, marketed in the hope of reducing the peak plasma lithium level and thereby the concomitant side effects as well, should be taken with a grain of soluble salt. The patient who complains of nausea or mild gastric acidity a few hours after taking lithium would be well advised to simply divide his daily dose into two, taken

R. H. Belmaker, P. Lichtenberg, *Psychopharmacology Reconsidered*, https://doi.org/10.1007/978-3-031-40371-2_8

a few hours apart. Alternatively, one may take the full daily dose before going to bed, which some small studies have found to cause fewer side effects without impairing the clinical benefits. Nevertheless, in light of the aforementioned requirement for the hydrophilic lithium ion to cross the cell membrane into the cytoplasm at therapeutic levels in order to have an effect, it is pharmacologically counterintuitive that allowing low levels most of the day is equally beneficial. Most clinicians will recommend dividing the daily dose into two or even three portions, so as to minimize side effects while maintaining stable intracellular blood levels. Intravenous administration of lithium is forbidden, lest it result in life-threatening cardiac arrhythmias.

Lithium will accumulate gradually in plasma and enter the intracellular space, reaching after approximately 10 days a state of equilibrium. In the presence of normally functioning kidneys, the half-life of lithium is about a day. Dehydration, hyponatremia and other conditions can decrease lithium excretion markedly, leading to potentially toxic plasma lithium levels. After 10 days, lithium levels in the plasma should be tested. Blood level should be determined in the morning before the first daily lithium dose as these are felt to be most predictive of intracellular levels. Testing before 10 days of treatment could potentially mislead the clinician into thinking that a therapeutic level has been obtained, while in reality the level may continue to rise in days that follow, with possibly toxic implications. This understanding of the pharmacokinetics of lithium is therefore very useful for the prescribing psychiatrist. Rarely should the dose be adjusted at more frequent intervals, with the exception of the start of lithium treatment, when intentionally subtherapeutic doses are offered so as to establish patient tolerance and acceptance.

The therapeutic effect of lithium in the treatment of acute mania will generally require 10 days, which likely is a result of the time necessary to achieve steady state cytoplasmic levels. This is somewhat briefer then the 3 weeks often reported before the onset of treatment effects when taking antidepressants or antipsychotics.

As might be anticipated from the pharmacodynamics of lithium, by 2 to 4 weeks prophylactic treatment for bipolar disorder will reduce the rate of relapse, compared with a control group. Discontinuing lithium after prolonged prophylaxis may lead to some rebound exacerbation, again compared with a control group. If discontinuation is necessary, it is best done gradually, over weeks or even months, while a different mood stabilizer can be introduced simultaneously.

A lithium overdose, which is a medical emergency, can be pharmacokinetically confusing. The patient might present at the emergency room after swallowing a huge number of tablets, and found to have severely toxic blood levels as high as 10 mM/liter, and yet appear asymptomatic on examination. If the physician mistakenly sends this patient home, he will likely return by next morning in a comatose state. As the reader will by now understand, this situation is a result of the time lag of many hours from gastric absorption into the blood stream until the lithium enters the brain cells where it will exert its toxic effects.

Conversely, the clinician must also be alert to the case of the patient treated with lithium, at stable blood levels, who continues treatment while not bothering to follow up on his blood levels, maintaining the same dose for years as he and his kidneys age. Perhaps he starts a low sodium diet, or increases his lithium dose to combat a bout of low mood. This person will be at risk for developing lithium toxicity with severe symptoms (see below) such as diarrhea and seizures even with a blood level of 2.5 mM, since the cytoplasmic level will have had ample opportunity to attain full equilibrium with the plasma level.

Natural variation in renal clearance of lithium amongst different people, and even the same person in different circumstances, coupled with a narrow therapeutic index, require monitoring lithium blood levels. Therapeutic doses will therefore vary amongst patients. Target blood levels in lithium treatment are 0.5–1.0 mM. Where lithium monotherapy is offered for acute mania, levels of 1.0–1.5 might be even more beneficial. For prophylaxis, levels of 0.6–0.8 mM have been reported to be more effective than levels of 0.4–0.6 mM, but levels of 0.4–0.6 will cause fewer side effects than the higher levels. Therefore, there is no one right level, but rather the clinician must weigh the pros and cons for each patient, who should be a partner in the deliberation.

Some have reported that for bipolar depression, which can be difficult to treat, lithium levels of 1.0–1.5 mM may be more effective as an antidepressant; however, good controlled data is lacking. Moreover, in depression, poor oral hydration and undernutrition leading to sodium depletion demand even more rigorous monitoring of lithium levels.

Efficacy of Lithium in Bipolar Disorder

Schou's classic studies of lithium have been largely confirmed. Lithium is efficacious in acute mania, in preventing relapses of mania and depression, and to a lesser degree for the treatment of acute bipolar depression [1].

Research regarding the therapeutic effects of lithium was performed before the year 2000 without funding by the pharmaceutical industry. This body of research has been criticized for falling short of FDA standards requiring larger numbers of subjects, rigorous quantitative rating scales with predetermined endpoints, and firm assurances that double blindness was maintained. These criticisms of the foundational studies of lithium in bipolar disorder served commercial interests for marketing newer, patentable drugs claiming efficacy in bipolar disorder.

Carbamazepine

After lithium's acceptance, Okuma, a Japanese researcher, serendipitously discovered that carbamazepine is also helpful in bipolar disorder. Carbamazepine is an antiepileptic compound unrelated structurally to lithium, but with chemical similarities to chlorpromazine.. Okuma's work stimulated many further studies by Post and others, who postulated an epilepsy-like etiology of bipolar disorder and a kindling model of the course of bipolar disorder. These theories remain speculative: lithium itself is devoid of anti-epileptic effect, while not all anti-epileptics are effective in bipolar disorder [2].

Carbamazepine requires therapeutic drug monitoring but has a wider therapeutic index than lithium. Overdose in less dangerous; the initial signs involve ataxia and diplopia. Long term treatment can cause liver toxicity. It has been less studied because of lack of commercial interest but controlled studies are convincing although small. There is no association of response with EEG abnormalities although that was an early exciting hypothesis. Carbamazepine can be used together with lithium in patients who are not adequately responsive to either; this suggests a different mechanism of action (Contrast this with dopamine receptor blockers, where most evidence suggests that polypharmacy of two dopamine receptor blockers is futile).

Valproic Acid

The next drug found efficacious in bipolar disorder, this time by Emrich in Germany, was valproic acid, a 5-carbon branched chain carboxylic acid which also bears no chemical relation to lithium, though like the latter it may be present naturally in geological samples. The pharmaceutical industry energetically promoted valproic acid for bipolar disorder, motivated by a special patent obtained for divalproex sodium (which dissolves into valproate in the stomach). For this round of studies over the past twenty years in acute mania and the prophylaxis of bipolar disorder, lithium was used as a positive control. These large, modern pharmaceutical-funded studies provided good data for the use of lithium in bipolar disorder. Lithium proved neither better nor worse than valproate. While some claimed to have found clinical markers for a specific Li response, these results were not generally replicable. Some have claimed that valproate is better than lithium for mixed affective episodes, or for dysphoric mania, or for rapid cycling bipolar disorder, but none of these predictors has replicated to a convincing degree. Open minded clinical trial and error over time is probably the best strategy today. All these drugs are also efficacious in acute mania as well as in the prophylaxis of both the manic and depressive poles in bipolar disorder, and perhaps less helpful in the treatment of bipolar depression [3].

Patients who have responded to but cannot tolerate valproate side effects can be switched to carbamazepine, and vice versa. The same holds for lithium and valproic acid, as well as for valproic acid and carbamazepine [4]. Valproate can be combined with carbamazepine or with lithium and all three can be combined in rare difficult patients to achieve a response. No one has been able to establish the profile of the patient who clearly responds only to lithium and not to valproic acid, or who responds to both but not to carbamazepine (see Fig. 8.1).

A newer anti-epileptic compound of interest is lamotrigine, which seems useful for treating and preventing bipolar depression, but shows little or no efficacy in acute mania, and uncertain efficacy for the prevention of mania. Lamotrigine too is

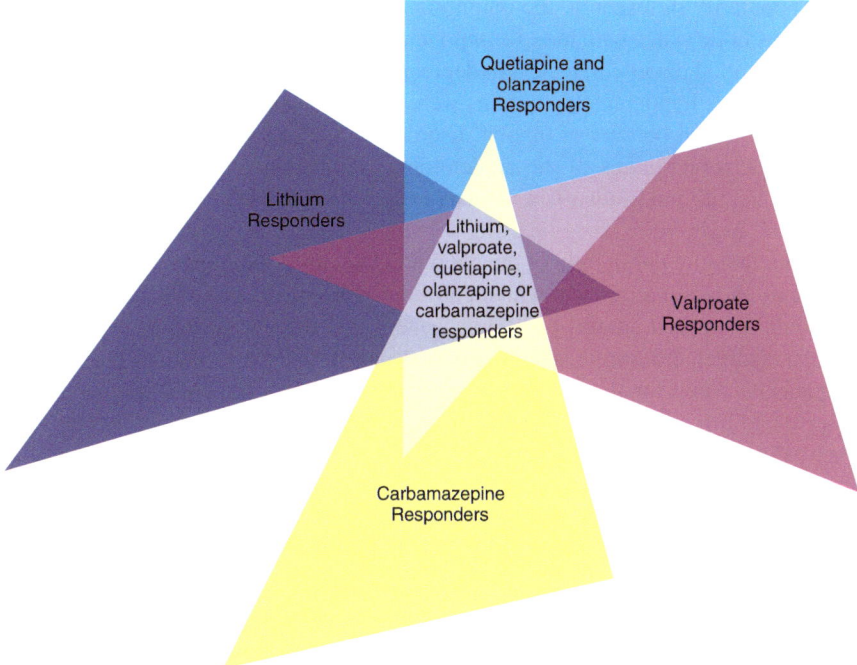

Fig. 8.1 The overlap of mood stabilizer clinical effects

chemically unrelated to the other lithium-like medications we have described. The efficacy of lithium in bipolar disorder was reconfirmed in studies of lamotrigine, where lithium served as a positive control in large pharmaceutical-sponsored multi-center trials.

Mood Stabilizers

The group of medications we have been discussing are often termed "mood stabilizers". One must keep in mind that this does not imply that they affect people's daily or hourly variability of mood. In addition, there is no evidence that they benefit the mood instability common in borderline personality disorder. Their effect upon the mood in unipolar depressive disorder or dysthymic disorder is limited or non-existent. Patients and their families may misunderstand the term "mood stabilizer" in psychiatric jargon. The psychiatrist may need to clarify that these medications are indicated only for the extreme mood swings which occur in bipolar disorder, much as one would give aspirin to reduce body temperature only in the presence of a fever.

As will occasionally happen in psychopharmacology, the apparent success of a class of medications for a particular diagnosis will lead to an expansion in the situations where that diagnosis will be applied. Following lithium's clinical success and

early popularity in research, the diagnosis of manic depressive illness was broadened. Psychotic states with mood components were included in the DSM-4 and 5 as
psychotic mood disorders; recurrent depressions with occasional episodes of hypo-
mania or some hypomanic symptoms were rebranded as bipolar II. This widening
of the conceptual borders of bipolar disorder also led to a concomitant expansion
in\\of population studies of the epidemiological incidence and prevalence of the
disorder. Yet lithium is not effective in most cases of bipolar II disorder. The treating
physician needs to be wary of using lithium in situations where the benefits are
doubtful, in particular in light of the ongoing blood level monitoring it requires, and
the short- and long-term risks which consuming this medication entails [5].

While controlled studies of lithium have found it to be superior to placebo in
bipolar disorder, these studies represent as always the average result for large groups
of patients, while for the patient sitting opposite you in the consulting room the
response to lithium can vary greatly. Responsiveness to lithium is not co-extensive
with a diagnosis of bipolar disorder, regardless of which edition of DSM or ICD we
use to determine the diagnosis. Some we are clearly dealing with bipolar disorder
will not respond to lithium, or will respond only partially in terms of frequency and
severity of episodes. It is best to avoid unrealistically raising expectations in newly
diagnosed patients and their families. Many patients respond, yet others do not. For
the non- and partial-responders, other options are available. Continued follow-up
and monitoring by the psychiatrist is essential. Conversely, for patients who fail to
achieve good remission with other mood stabilizers, switching to lithium is an
option [6].

Many considered lithium "the gold standard of treatment of bipolar disorders"
and the "first-line" treatment for bipolar disorder. These are misrepresentative state-
ments. Not all newly diagnosed bipolar patients should necessarily be offered treatment with lithium and no guideline makes such a recommendation. Lithium
treatment entails regular blood testing, dietary restrictions, stable physical health,
and the good fortune of being a responder. Lithium is the classic therapy for bipolar
disorder, but not the gold standard. Lithium is clearly not for every patient [7].

Antipsychotics as Mood Stabilizers

After the discovery of clozapine as a uniquely effective antipsychotic in schizophrenia without inducing EPS, a huge effort began to find a clozapine-like antipsychotic
without the clozapine side effect of agranulocytosis. The development of olanzapine, risperidone and quetiapine in the early 2000s as antipsychotics with reduced
EPS encouraged their trial in various phases of bipolar disorder. It was already well
known that first generation dopamine D-2 blocker antipsychotics were effective in
mania. The new compounds, often with 5- HT2 blocking properties as well as D-2
blockade, were found effective in prophylaxis of mania, prophylaxis of bipolar
depression, and in many cases in treatment of bipolar depression. They may be more
effective than lithium in treatment of bipolar depression, perhaps since they seem to

have clear anxiolytic and hypnotic effects. Since the 2000s almost every new anti-psychotic developed appears to be effective in bipolar disorder: not all have received FDA indications but perhaps only because of technically failed studies or lack of commercial pharmaceutical company interest to develop this indication. Is this real progress? For some patients, definitely yes. As shown in Fig. 8.1, atypical antipsychotics help some patients that do not respond to lithium or anti-epileptic mood stabilizers. Is it a theoretical breakthrough? Yes and no. Early textbooks of psychopharmacology emphasized that chlorpromazine helped in mania, schizophrenia and depression and was not diagnosis specific. Some historical evidence suggests that fluphenazine decanoate was widely used in prophylaxis of bipolar disorder even if the patients were diagnosed formally as suffering from schizophrenia. So second generation antipsychotics may not be a breakthrough in bipolar disorder; merely an excuse to return to early use of D-2 blockade as a mechanism to ameliorate bipolar disorder. This further complicates efforts to find a common mechanism of action for treatment of bipolar disorder. But it is better to live with and accept ambiguity than to believe simple and clear dogmatic untruths. Kishi et al. have used a new and elegant statistical design called network meta-analysis. The usual meta-analysis combines many studies comparing two drugs to reach an overall average conclusion with stronger certainty. Network meta-analysis compares studies of Drug A vs Drug B with studies of Drug B vs Drug C with studies of Drug C vs. Drug D, etc. It is clearly shown that lithium, carbamazepine, valproate, olanzapine, and quetiapine are no better or worse than each other for the average patient [8, 9]. Individual differences are a different matter.

While largely avoiding EPS, second generation antipsychotics are very problematic with their side effects of weight gain, hyperlipidemia and hyperglycemia [10]. The mixed dopamine agonist-antagonist compounds developed for schizophrenia are closer to weight neutral and metabolically safe: they appear to also be mood stabilizers although each should be examined carefully. The preferentially D-3 antagonist compounds such as cariprazine and brexpiprazol also seem to be effective mood stabilizers although the prototypical aripiprazole was useful in depression but rarely in mania.

Administration of Lithium

Before initiating lithium treatment, tests should be ordered for blood chemistry (including renal function and calcium), a complete blood count, and TSH, as well as an EKG in patients over 60. A recent physical and neurological examination should be documented. If these have not already been performed, they can be briefly done by the psychiatrist, so that the start of treatment not be inordinately delayed.

In contemporary psychiatric practice, lithium is rarely administered alone, for several reasons. First of all, many public mental health care services are pressured by insurance carriers, in the United States and elsewhere, to reduce duration of treatment, in particular expensive inpatient stays. Antipsychotic treatment can produce more rapid behavioral results, with lithium added for prophylaxis while in the community. Under these circumstances, it is advisable to start lithium in the acute hospital phase only when the patient can make a convincing commitment to continue using it in the future. Sudden discontinuation of lithium should be avoided, because of the danger of a rebound relapse. Secondly, lithium treatment requires the patient's cooperation with blood level testing, nutrition and hydration; this cooperation will not always be forthcoming. Thirdly, lithium monotherapy used as prophylaxis is also rare nowadays, because many patients will have a psychotic component to their bipolar episodes, requiring the use of an antipsychotic with their lithium. Should side effects of lithium develop, many psychiatrists will prefer to replace it with a second generation antipsychotic not requiring blood testing, thus avoiding the medico-legal risk of prescribing lithium without blood monitoring [11].

The Importance of Patient History

Since the guidelines for the treatment of an acute manic episode generally advise prescribing a mood stabilizer such as lithium, carbamazepine, valproate, olanzapine, quetiapine or lurasidone, a psychiatric first responder in an emergency room, overwhelmed by the workload and faced with a manic patient, might choose one of these medications at random. Where possible, this should be avoided. Most people requiring treatment in an acute manic state are veterans of previous episodes and treatment; the results of these earlier treatments should be assiduously sought by the clinician before deciding which treatment to implement now. Computerized records should be carefully read, and family members patiently queried. Ask the patient, repeatedly if necessary, what he remembers, and whether he might be carrying any previous medical records on his person, or know of such records at home, where perhaps a family member could locate them and report by phone on their contents. **Patient history is the best, most reliable predictor of patient response** [12]. One who in the past has responded to valproic acid, for example, should not be started on lithium on the misleading assumption of its therapeutic equivalency. Another patient who has done well with lithium for years, but after discontinuing lithium is in the throes of a manic attack, should not be switched to carbamazepine, regardless of what may be written in the latest guidelines about drug equivalence.

Researchers have understandably searched for the shared biochemical effects of lithium, valproate and carbamazepine, all considered "mood stabilizers", which might explain their therapeutic effects. Manji and colleagues [13, 14] in particular have followed this line of inquiry. But all we can say is that no common dominator has been found that explains the biochemical basis for calling these three compounds mood stabilizers. To complicate the situation further, data has accumulated that many second generation antipsychotics are effective and convenient both in preventing mania and preventing depression, so that they too can reasonably be called mood stabilizers. These medications share the mechanism of dopamine blockade, which is clearly not a primary effect of lithium. We must honestly admit that the answer to the textbook question of how lithium works is that we simply don't know [15].

Suicidality

Intriguingly, treatment with lithium is correlated with a reduction in both serious suicide attempts and completed suicide [16]. Since bipolar disorder is associated with high suicide rates, the ability of lithium to prevent suicide is potentially very important. Some researchers have considered suicidality to be a psychiatric syndrome in its own right, with its own pathophysiological mechanisms and treatment requirements, unrelated to the disorder with which it may occurs. Unfortunately, evidence of lithium's anti-suicidal effects apart from the context of bipolar disorder is not sufficiently persuasive to recommend lithium in the treatment of suicidal patients without bipolar disorder, such as recurrently suicidal borderline patients. But for patients with bipolar disorder, when considering various treatment options, certainly the evidence for lithium's anti-suicidal action ought to be taken into account [17].

No mechanism for lithium's anti-suicide effect is known. It is worth noting, however, that clozapine has also been shown to reduce suicide, beyond the effect of other antipsychotic drugs. Common to both lithium and clozapine is that they require frequent blood level monitoring. A tantalizing possibility is that the added attention of mental health care workers, as well as the discipline of the patient who is required to agree to regular blood tests, are the real reason why these two compounds correlate with anti-suicidal effects [16].

Individual Differences and Prediction of Lithium Response in Bipolar Disorder

Repeated attempts over the years by clinicians and researchers have yielded no replicable clinical syndrome to be strongly predictive of lithium response in bipolar disorder. Psychosis in the manic depressive phase has not been shown to be an absolute negative predictor. A family history positive for bipolar disorder is also not a strong predictor of response, though a family history of response to lithium has not been ruled out as a predictor of response, and should be considered accordingly by the clinician. A positive family history of valproate response or carbamazepine response does not preclude a lithium response in the individual patient. While many studies have sought to identify genomic variants by GWAS that might predict lithium response, none of these have proven useful at a clinical level, notwithstanding efforts by several companies to market directly to patient consumers genetic tests for lithium response [18–20].

As noted earlier in the chapter, in controlled studies, statistical analyses confirm that lithium is superior to placebo for the treatment of mania and for preventing further episodes. A closer appraisal of the results, however, will show, as is so often the case in medicine in general, a huge variation in response. Roughly one-third of patients will respond to placebo, two-thirds will respond to lithium, and the last third will not respond even to lithium. To render matters even more complex, some individuals improve completely, others only partly, and yet others in only one symptom while the wider disease syndrome remains. Our ability, then, to predict the response of the individual patient under our care remains a major concern in psychiatry, as it is elsewhere in modern medicine. The goal is to achieve a level of "precision medicine" wherein drug treatments are precisely "fitted" to the patient who will respond. This has not been achieved for lithium treatment, nor for that matter for most treatments in medicine. One might say that this highlights the distinction between the scientific method, which seeks to discern principles, the more generalizable the better, and the approach of the physician, who wants first of all to know what will help this particular patient seeking our care. For Galileo rolling metal balls down an incline, minute variations in final acceleration that were due to tiny differences in the weights of the balls, or in the force of the ambient wind, or in friction with the surface of the incline, could be dismissed as "noise". Truth was the physical law derived from the mean of multiple measurements. In medicine, on the other hand, including psychopharmacology, the truth mainly interests us in our daily labors is not the mean, but rather the individual patient whom we are trying to help. For this purpose, the controlled clinical trial can offer at most only a gross indicator of what might be the best treatment for the patient in our care—and that too only on the condition that the actual patient in the consulting room would fit precisely the inclusion criteria for the trial, which is often not the case More compelling still is the fact that a research trial's inclusion criteria do not necessarily define a recognized clinical entity, but may rather include a heterogenous disorder or only part of the true pharmacological target of the treatment. The clinician must understand that a trial that shows that lithium, for example, is effective in DSM-V bipolar disorder, does not indicate that the DSM-V-defined nosological entity of bipolar disorder is coextensive with the actual boundaries of lithium response.

Toxicity

Lithium, as noted, has a very narrow therapeutic index, with therapeutic plasma levels from 0.5 mM to 0.8 mM. Moreover, lithium bears a high potential for severe toxicity: levels above 1.5 mM in the plasma will produce toxic effects, likely beginning with polyuria, abdominal pain, vomiting and diarrhea. Since lithium is not excreted in vomitus, diarrheal stool or in the dilute urine of lithium nephrotoxicity, these side effects will themselves further increase lithium plasma levels, setting up a potentially lethal vicious cycle. By this mechanism, once it begins, lithium toxicity is liable to advance rapidly. At blood levels above 2.0 mM, cardiac abnormalities are common and grand mal seizures frequently occur. Mortality will be high. Lithium inhibits numerous magnesium dependent enzymes, yet intravenous magnesium sulfate is ineffective in treating lithium toxicity. Dialysis is often implemented, but with a gradient of lithium at 2 to 3 mM in the blood, the dialysis may not be efficient. When possible, in the presence of lithium toxicity, oral rehydration with liquids containing salt should be attempted. Where that is not an option, intravenous rehydration with isotonic saline is necessary. Renal function must be carefully monitored [21].

The key to preventing lithium toxicity begins with patient education. Each patient should be provided with written information detailing the common side effects that occur with lithium even at therapeutic doses, and—as a separate list—symptoms resulting from early toxicity. Additional information necessary to convey to the patient, preferably in written form, should include the crucial importance of maintaining normal hydration, avoiding low-sodium diets, and informing the treating psychiatrist if the patient starts taking a new medication such as a thiazide diuretic. The most common contemporary cause of lithium toxicity is the concomitant use of a thiazide anti-hypertensive medication prescribed by a family physician who is unaware that the patient is receiving lithium. Many NSAIDs also can impair lithium excretion and induce toxicity, so patients must be aware of this danger as well, and if NSAIDs are necessary, the patient must have his blood levels monitored. The medical profession and indeed the general public would do well to relate to lithium the way they do to digitalis: as an extremely useful compound both in the history of medicine and in contemporary practice, but which by the nature of its effect upon basic physiological systems has the capacity to induce severe and potentially fatal toxicity.

Side Effects

The possible side effects of lithium treatment are myriad, and include many which may be unfamiliar to the clinical psychiatrist who does not prescribe lithium regularly. The physician must familiarize himself with the side effects of lithium, and discuss them fully with the patient embarking on such treatment. Since some of lithium's serious side effects impinge upon physiological systems outside of the central nervous system, where hormone receptors and second messenger systems are involved, it was therefore believed for a time that lithium's side effects involving the thyroid, parathyroid and kidney may also involve second messengers in these tissues. This has not turned out to be the case, and the mechanism for lithium's side effects in each of these remains complex and not fully understood.

Polyuria and Polydipsia

Lithium causes a reversible nephrogenic diabetes insipidus by blockading the effect of antidiuretic hormone (ADH) in the collecting ducts of the kidney, so that the kidneys cannot concentrate the urine. This is the mechanism underlying polyuria and polydipsia, the most common renal side effect of lithium. The patient should understand that this is in no way a sign of underlying renal disease, and should not be confused with the rarer but far more serious nephrotoxicity of lithium which he may have read about on the internet [22].

Parallel to TSH in the thyroid, lithium at therapeutic levels will not have an effect on the ADH receptor itself. The blockade of the ADH receptor probably occurs distal to the receptor, but is much more complex than early formulations of effects on the second messenger system had led us to understand. The blockade of the antidiuretic hormone produces a diuresis of dilute urine. Almost all patients starting lithium treatment will experience a polydipsia (thirst leading to increased drinking) and polyuria (increased necessity for urination). The physician needs to alert her patient not to satisfy his thirst with sweet drinks, which is liable to lead to weight gain which is not directly caused by lithium itself, although the lithium is commonly blamed for this. The psychiatrist should similarly warn her patient to anticipate increased urination, especially in the presence of benign prostatic hyperplasia (BPH) or other possible causes of polyuria and nocturia. Unfortunately, this lithium-induced polyuria-polydipsia will not be responsive clinically or experimentally to antidiuretic hormone, neither by injection nor in the sublingual or intranasal spray forms (DDAVP) which are usually employed when treating primary enuresis in children or adults before bed or used to treat pituitary derived forms of diabetes insipidus. The reason for this, as noted above, is that the block induced by lithium occurs distal to the receptor.

The polyuria and polydipsia induced by lithium may with time diminish or even disappear, or remain merely a nuisance with no actual health risk, for many months or even years, only to return with renewed severity during the protracted period of prophylactic treatment with lithium. Why these fluctuations and re-exacerbations occur are often unknown. Reducing the lithium dose while monitoring blood levels can help.

For reasons only partially understood and beyond the purview of this textbook, thiazide diuretics and the potassium sparing diuretic amiloride are paradoxically quite useful in the relief of lithium-induced polyuria and polydipsia. However, the treating psychiatrist must realize that the thiazide treatment of lithium induced polyuria-polydipsia with thiazide can raise lithium levels, which therefore must be carefully monitored. The psychiatrist is a physician and as such, ought not to hesitate to supervise himself the thiazide treatment, much as we recommended above regarding the treatment of lithium-induced hypothyroidism with thyroxine. If the psychiatrist opts to refer to a nephrologist, the risk is too high that the latter will advise discontinuing lithium, which is often not actually necessary. If the nature of the problem makes a nephrological consultation unavoidable, the psychiatrist should at the least choose a nephrologist who is understands the psychiatric issues involved, and is amenable to cooperation with the psychiatrist.

Tremor

Most patients receiving lithium will develop a mild resting tremor, which should be observed and followed regularly by the psychiatrist. Do not confuse this with a Parkinsonian tremor, as the patient will not present with akinesia, mask facies, festinating gait, or cogwheel rigidity, unless they happen to be simultaneously taking neuroleptics. The tremor of lithium is more akin to what has been called a MacDonald-Critchley essential tremor. Once again, beware of referral to a another specialist, in this case a neurologist, lest the recommendation arrive that the patient discontinue lithium, which would be an error. The essential tremor if lithium requires neither neurological consultation nor—unless unusually severe, or not improved by dosage reduction, or in the case of a person whose livelihood and/or self-esteem depend on steady hands—discontinuation of lithium. If necessary, it can often be adequately treated with a small dose of propranolol, such as 10 mg three times daily. Propranolol carries its own side effects, and should be added only if the tremor causes more than a slight inconvenience. Anticholinergic anti-Parkinson medication will be of no benefit.

Mood Stabilizers in Schizoaffective Disorder

Is lithium relevant for the treatment of people diagnosed with schizoaffective disorder? Two different approaches which have developed in regard to this question raise troubling questions about how knowledge is sought and consensus is reached in the psychiatric profession.

In the 1970s, when people who developed psychosis in the midst of manic episodes tended to be diagnosed as with schizophrenia, clinicians in the USA were excited to find a role for lithium in the treatment of these episodes. Nosological mores may follow therapeutic interventions, and the bipolar disorder came to encompass people with mood-congruent delusions and hallucinations in both depression and mania and, subsequently, even those with mood incongruent delusions and hallucinations. Following these changes, studies of bipolar disorder today find than more than one third deteriorate cognitively while more than 80% receive disability insurance. This is the profile of people who used to be diagnosed with schizophrenia. Moreover, many clinical case reports documented a responses to lithium in patients diagnosed with schizophrenia, even those with its chronic forms, particularly when they suffered from recurrent manic or depressive episodes. A Jerusalem study at the time [23] examined "excited psychosis" in sequential psychiatric admissions. All patients received haloperidol, and in addition randomly received either lithium or placebo. The study found evidence for a significant benefit for those receiving lithium. This applied for all subjects, regardless of whether they met the criteria for non-psychotic mania, psychotic mania, schizoaffective disorder or chronic schizophrenia with exacerbation, though the response was attenuated as patients were placed further along the continuum between non-psychotic mania and schizophrenia. But all in all, in this approach, lithium plays a therapeutic role wherever affective symptoms are present in psychosis.

And yet, a different, completely contradictory tradition exists as well in psychiatry. According to this second approach, one should expect a therapeutic response to lithium in the non-psychotic bipolar patient presenting in a manic state characterized by euphoria (rather than irritability, which today is more common), in the absence of comorbidity (such as borderline personality, anxiety disorders, or substance abuse that characterize most people diagnosed today with bipolar disorder) and with a non-deteriorating life course, no cognitive impairment, and the attainment of a full remission between each episode. These clinical characteristics, according to this tradition, predict lithium response.

Reading these differing approaches, it can be difficult to see how they both sprouted from within the same profession, presumably working with the same clinical material. Perhaps psychiatrists, like other physicians, develop schemas in their mind based on a the first textbook that they read or the first manic patient they met. Our opinion is that based on a sober assessment of the evidence, lithium is clearly valuable not only for treating the classic manic patient, but also for one meeting diagnostic criteria for schizoaffective or even schizophrenic spectrum disorder, in the presence of an affective component to their episodes, and who have not responded satisfactorily to other therapies.

Mood Stabilizers in Unipolar Depression Maintenance Therapy

Lithium has not proven effective as monotherapy maintenance treatment for the prevention of unipolar depressive disorder, which occurs at least once in the lifetime of perhaps 20% of the general population, as defined by the broadly inclusive contemporary DSM criteria. This contradicts images in the popular media, where lithium is often portrayed as the ultimate antidepressant. This message can be dangerously misleading, by contributing to the over = prescription of lithium, with its concomitant side effects, as discussed above. Yet in considering this issue, we would do well to remember the huge expansion of the concept of depression during the past half century. In 1969, in Klein and Davis's textbook of psychopharmacology, depression could still be considered a "rare disease". We do not have good data specifically for that earlier, more severe conception of depressive illness. Perhaps in those cultures, or countries where depressive disorder is still narrowly defined as an endogenous, severe illness, some may better respond to lithium prophylaxis.

Moreover, some evidence suggests that a family history of bipolar disorder responding to lithium can predict a role for lithium prophylaxis preventing recurrent unipolar depressive episodes.

The overuse of lithium in unipolar depressive disorder may be as troubling as its underuse in bipolar disorder. As is too common in medicine, it can be difficult to estimate the extent to which medicines benefit and the extent to which they harm. One is again reminded of the old Osler adage, which unfortunately remains apposite today: if all the medicines in the world were thrown into the sea, it would be very good for mankind but very bad for the fishes.

Is Lithium in any Sense Specific?

Lithium is effective against the euphoric mood in manic episodes and the depressive mood in bipolar depression; much less is known about its effect on mood in normal volunteers though such effect, if present, is certainly less dramatic than what one expects in the presence of bipolar disorder. In this sense lithium is "specific". However, if the claim of specificity is that lithium helps only and always in bipolar disorder, then as we have discussed above, this is simply not the case. Lithium can also be useful for many people diagnosed with schizoaffective disorder, as augmentation of the response to antidepressants in unipolar disorder, for episodic aggression not in the context of mood disorders, and in cluster headaches [24]. The concept of lithium specificity once bore heuristic value, generating much research over the past 70 years. This is no longer the case, and the claim of lithium specificity has outlived its relevance. The treatment options for mania, for bipolar prophylaxis, and for bipolar depression are hardly limited to lithium, which in its turn can be useful for conditions unrelated to bipolar disorder.

Clinical Vignettes

1. Garcia was a 33 year old engineer who developed a depression after a loss in the stock market. He suffered crying spells, insomnia and weight loss and did not work for 2 months. The depression resolved with no treatment. A year later he developed a mania with irritability, excessive spending, sexual improprieties and grandiose delusions. He responded well to lithium and continued on lithium prophylaxis with no recurrence of affective disorder for 5 years. With no apparent cause he developed then intolerable polyuria and the lithium was seamlessly replaced with carbamazepine. Garcia was well on carbamazepine for another 2 years until his liver enzymes began to rise such that the carbamazepine was stopped. He disappeared from follow up but apparently was well for 3 years until he returned in a manic episode. The symptoms were similar to his first manic

episode. He was treated with valproic acid and has been well affectively for over a decade with no side effects.
2. Avi had a manic attack at age 18 and on recovery developed a chronic depression. After almost 2 years of depression, he responded to olanzapine treatment and was then well for over 5 years. He gained weight excessively during the olanzapine treatment but attempts to wean him off olanzapine and on to valproate, carbamazepine or lithium were unsuccessful because of signs of affective relapse. He is now a candidate for cariprazine treatment.
3. Sara is 55 year old woman who had a manic attack postpartum at age 27. Her late mother was a lithium responsive bipolar patient. Sara was put on lithium maintenance at age 27 and has been well for the subsequent 28 years.

References

1. Geddes JR, Burgess S, Hawton K, Jamison K, Goodwin GM. Long-term lithium therapy for bipolar disorder: systematic review and meta-analysis of randomized controlled trials. Am J Psychiatry. 2004;161(2):217–22.
2. McIntyre RS, Berk M, Brietzke E, Goldstein BI, López-Jaramillo C, Kessing LV, et al. Bipolar disorders. Lancet. 2020;396(10265):1841–56.
3. Geddes JR, Goodwin GM, Rendell J, Azorin JM, Cipriani A, Ostacher MJ, et al. Lithium plus valproate combination therapy versus monotherapy for relapse prevention in bipolar I disorder (BALANCE): a randomised open-label trial. Lancet. 2010;375(9712):385–95.
4. Cipriani A, Barbui C, Salanti G, Rendell J, Brown R, Stockton S, et al. Comparative efficacy and acceptability of antimanic drugs in acute mania: a multiple-treatments meta-analysis. Lancet. 2011;378(9799):1306–15.
5. Young AH, Hammond JM. Lithium in mood disorders: increasing evidence base, declining use? Br J Psychiatry. 2007;191:474–6.
6. Miura T, Noma H, Furukawa TA, Mitsuyasu H, Tanaka S, Stockton S, et al. Comparative efficacy and tolerability of pharmacological treatments in the maintenance treatment of bipolar disorder: a systematic review and network meta-analysis. Lancet Psychiatry. 2014;1(5):351–9.
7. Yatham LN, Kennedy SH, Parikh SV, Schaffer A, Bond DJ, Frey BN, et al. Canadian network for mood and anxiety treatments (CANMAT) and International Society for Bipolar Disorders (ISBD) 2018 guidelines for the management of patients with bipolar disorder. Bipolar Disord. 2018;20(2):97–170.
8. Kishi T, Ikuta T, Matsuda Y, Sakuma K, Okuya M, Mishima K, et al. Mood stabilizers and/or antipsychotics for bipolar disorder in the maintenance phase: a systematic review and network meta-analysis of randomized controlled trials. Mol Psychiatry. 2021;26(8):4146–57.
9. Kishi T, Matsuda Y, Sakuma K, Okuya M, Mishima K, Iwata N. Recurrence rates in stable bipolar disorder patients after drug discontinuation v. drug maintenance: a systematic review and meta-analysis. Psychol Med. 2021;51(15):2721–9.
10. Belmaker RH. Lurasidone and bipolar disorder. Am J Psychiatry. 2014;171(2):131–3.
11. Lin Y, Mojtabai R, Goes FS, Zandi PP. Trends in prescriptions of lithium and other medications for patients with bipolar disorder in office-based practices in the United States: 1996-2015. J Affect Disord. 2020;276:883–9.
12. Belmaker R, Bersudsky Y, Agam G. Individual differences and evidence-based psychopharmacology. BMC Med. 2012;10:110.
13. Manji H, Bowden C, Belmaker R. Bipolar medications: mechanisms of action. APA Press; 2000.

14. Belmaker RH. Treatment of bipolar depression. N Engl J Med. 2007;356(17):1771–3.
15. Belmaker RH. Patient history must be incorporated into any guidelines. Bipolar Disord. 2009;11(7):772; author reply 3, 772.
16. Smith KA, Cipriani A. Lithium and suicide in mood disorders: updated meta-review of the scientific literature. Bipolar Disord. 2017;19(7):575–86.
17. Tondo L, Baldessarini RJ. Antisuicidal effects in mood disorders: are they unique to lithium? Pharmacopsychiatry. 2018;51(5):177–88.
18. Hou L, Heilbronner U, Degenhardt F, Adli M, Akiyama K, Akula N, et al. Genetic variants associated with response to lithium treatment in bipolar disorder: a genome-wide association study. Lancet. 2016;387(10023):1085–93.
19. Stern S, Santos R, Marchetto MC, Mendes APD, Rouleau GA, Biesmans S, et al. Neurons derived from patients with bipolar disorder divide into intrinsically different sub-populations of neurons, predicting the patients' responsiveness to lithium. Mol Psychiatry. 2018;23(6):1453–65.
20. Papiol S, Schulze TG, Alda M. Genetics of lithium response in bipolar disorder. Pharmacopsychiatry. 2018;51(5):206–11.
21. Decker BS, Goldfarb DS, Dargan PI, Friesen M, Gosselin S, Hoffman RS, et al. Extracorporeal treatment for lithium poisoning: systematic review and recommendations from the EXTRIP workgroup. Clin J Am Soc Nephrol. 2015;10(5):875–87.
22. Azab AN, Shnaider A, Osher Y, Wang D, Bersudsky Y, Belmaker RH. Lithium nephrotoxicity. Int J Bipolar Disord. 2015;3(1):28.
23. Biederman J, Lerner Y, Belmaker RH. Combination of lithium carbonate and haloperidol in schizo-affective disorder: a controlled study. Arch Gen Psychiatry. 1979;36(3):327–33.
24. Licht RW. Lithium: still a major option in the management of bipolar disorder. CNS Neurosci Ther. 2012;18(3):219–26.

Chapter 9
Electroconvulsive Therapy, Transcranial Magnetic Stimulation, Deep Brain Stimulation and tDCS

Electroconvulsive therapy (ECT), transcranial magnetic stimulation (TMS), deep brain stimulation (DBS) and transcranial direct current stimulation (tDCS) are four biological treatments in psychiatry that are not pharmacological but are traditionally covered in psychopharmacology texts because they are often alternatives to pharmacological treatment or serve in research studies as standards against which pharmacological treatments are compared. ECT was discovered by chance in the 1930s when it was believed that epileptic patients do not get schizophrenia and therefore Cerletti and Bini induced convulsions in patients with psychosis to try to ameliorate their symptoms. Of course, the opposite is now known to be the case since epileptic patients develop chronic psychosis at a much higher rate than the general population. Moreover, the new treatment of inducing convulsions turned out to help depressive patients far more than patients diagnosed with psychosis. The cynic may use this as an example of how little we know; or the optimist may see this as an example of how lucky the human race and medicine can sometimes be. ECT is a highly effective treatment of depression with the well-known drawbacks of expense, memory loss even if usually temporary, and high relapse rate.

ECT declined in use after the introduction of modern antidepressants and antipsychotics. However, it has not disappeared in psychiatric practice in almost any country because it remains an important tool for severely suicidal acute depressive patients, sometimes for acute and otherwise unmanageable manic patients, and occasionally for chronic treatment resistant depressive patients.

The induction of a convulsion by passing an electric current though the brain, originally from temple to temple, usually takes about a second using alternating current at voltages of about 100 volts. The ECT apparatus does not allow for electrocution because it is separated from the wall current source. After the 1 s stimulation, there is usually a 1 to 2 s refractory period and then the beginning of a classic tonic-clonic grand mal convulsion. The grand mal convulsion usually lasts for 20–60 s. In the postictal period, the patient has a flat EEG, no muscle movement and breathing returns only after a minute or two. Usually, the patient is completely

R. H. Belmaker, P. Lichtenberg, *Psychopharmacology Reconsidered*, https://doi.org/10.1007/978-3-031-40371-2_9

disoriented for 10–15 min and has memory impairment for up to 2–3 h after such a treatment. Since the 1960s, treatments are conducted almost universally under mild brief anaesthesia, so that the muscle movements of the grand mal convulsion are seen as slight twitches and the patient is unconscious for the whole process. Some have claimed that this use of anaesthesia prolongs the recovery phase and the disorientation following ECT.

For most of the history of ECT the grand mal convulsion was considered the essential therapeutic mechanism. Historically, convulsions induced by indoklon gas were reported to be equally effective but less safe. The treatments are usually given two to three times per week for a total of 8–12 treatments.

The clinical practice of ECT has been changed greatly by the comprehensive studies of Sackeim and his associates [1]. They compared effectiveness and memory impairment results in depressed patients whose ECT treatment was given only to the right side of the brain, only to the left side of the brain, through the whole brain, with low voltage electrical stimulation, with high voltage electrical stimulation, or with any of the combinations above. Over the years they also looked at variations of the usual 60 cycles alternating current electrical stimulus: for instance, cutting off the wave and only giving pulse stimulation. Their findings in brief summary indicate that low dose electrical stimulation of the right brain is less effective in depression even if a convulsion is induced but that pulse waves that give less total electrical stimulation but achieve a grand mal convulsion can be effective in depressed patients with less memory disturbance. Sackeim's work undermined the idea that the convulsion is the sine qua non of ECT because it seemed like convulsions induced by right brain electrical stimulation were ineffective. Moreover, it seems that the kinds of electricity used could affect the amount of memory side effect. ECT treatment based on Sackeim's findings is now widely used as standard practice, even though the apparatus has become much more expensive than it was in the past. EEG monitoring of convulsion length is now also part of standard practice because electroencephalographic evidence of convulsive process for at least 20 s does seem to be necessary for therapeutic benefit. Modern ECT machines adjust the voltage administered according to the electrical resistance across the skull of the particular patient, so that the dose is in current, milliamperes, rather than in volts. In less developed countries ECT is still usually administered with a small, inexpensive, box machine that is safe but has few options for different current, frequencies or electrical wave shapes. It is not clear if the wheel has been here reinvented or if this is indeed a better wheel.

It is difficult to do controlled clinical trials on ECT because of the severe nature of the illnesses for which it is recommended and because it is difficult to create a blind treatment. Almost all evidence that exists finds ECT very effective in acute depression and especially in acute suicidal depression. The side effects are almost entirely transitory including the memory side effects except for some long term "autobiographical memory" where certain periods of the patient's life may be erased from his memory [2]. This is rarely incapacitating in any way but must be emphasized in consent forms because its existence has been well proven in studies such by Squire and his group [3].

An acute electrically induced convulsion changes every neurochemical parameter that can be measured in the brain. It is therefore difficult to find the mechanism of action for ECT [4]. Anyone with a new biochemical assay will find some effect. The mechanism is not to cause the patient to forget his problems since there is little correlation between the degree of memory side effect and clinical effectiveness of ECT. Interestingly, ECT is also effective in mania and so it seems to be a mood stabilizer. It is not effective in antisocial personality disorder, in episodic aggressive personality disorder, in OCD unless accompanied by acute depression, in acute anxiety disorder, or in chronic schizophrenia.

The major clinical problem with ECT is the high relapse rate. It is a treatment that cannot be repeated frequently, and long-term antidepressant treatment must be instituted as soon as possible after the last in a series of ECT treatments when effective. An occasional rare patient that responds only to ECT and responds very well can be maintained for many years on one ECT treatment bi-weekly or monthly. These are the miracle cases that every psychiatrist remembers but they are the exception that proves the rule [5].

TMS

The work of Sackeim suggested that the electrical stimulation itself and not only the convulsion might be involved in the therapeutic properties of ECT. As early as 1904 following Maxwell's description of the electromagnetic concept it was suggested that magnetic stimulation could cause an electrical pulse in the brain. However, magnetic coils powerful enough to elicit an electrical stimulus of the brain were not yet available [6]. In the 1990s such coils became available, and it was soon shown that magnetic stimulation of the skull in a conscious patient could cause the movement of an arm, a leg or even a finger when the stimulation was performed above the appropriate area of the motor cortex. Grisaru et al. in 1994 [7] reported that such magnetic stimulation above the non-motor left frontal cortex had an antidepressant effect. Since then, considerable clinical research has taken place and magnetic stimulation of the brain has received some regulatory approval although the FDA only approved the device and does not demand full proof of efficacy for a device in the same way that it does for a drug. Many insurance companies do reimburse for TMS treatment for depression. Stimulation involves holding a coil containing electrical fibres above a defined point on the scalp. A rapidly alternating electrical current induces a collapsing and reforming magnetic field. Magnetic fields decline rapidly over distance and so the magnetic field penetrates perhaps only 1 cm into the brain. Where the magnetic field meets ions in neurons at the right angle, electrical fields are created that can discharge a neuron. Considerable publicity has been given to coils that are claimed to be configured to penetrate more deeply into brain and are called "Deep TMS". The evidence that these coils actually affect deeper areas of the brain is scant. "Deep TMS" should not be confused with Deep Brain Stimulation

(DBS, see below). It is highly unlikely that deeper areas of the limbic system are stimulated with currently available magnetic TMS coils [8].

Clinically, stimulation is usually carried out daily on weekdays for 2 to 4 weeks in sessions that last for 20 min. The guidelines for frequency and length of each session are not yet completely set. Excessive strength of magnetic field or length of treatment session can lead to a grand mal seizure which Lisanby and colleagues [9] have studied as "magnetic seizure therapy". Magnetic seizure therapy is experimental and not yet clinically used for any indication.

TMS is not equivalent to ECT and is not effective in acute suicidal depression and it is questionably effective in mania, perhaps dependant on which side of the brain is magnetically simulated. However, for depressive patients today who have failed treatment with one or two antidepressants, TMS is a viable option that helps more patients than placebo. It is effective in animal models of depression [10] but has no clear mechanism of action despite numerous reports of neurochemical effects.

Deep Brain Stimulation (DBS)

In severe late stages of Parkinson's Disease where L-DOPA is no longer effective and since the location in the brain of the deteriorated neurons causing Parkinson's is known, surgical implantation of an indwelling electrode has been shown to be clinically effective and is used routinely in a limited number of patients [11]. Some of these Parkinson's patients reported an improvement in mood. Therefore, implantation of an indwelling electrode has been tried in severe, resistant patients with depression and also in patients with resistant OCD. This treatment is still only experimental, and no proof of clinical indication exists. Numerous new potential sites of stimulation are becoming available due to research clinical findings in functional MRI and magneto-electroencephalographic studies, and some sites are inhibitory, others are excitatory, and whole circuits may be potentially affected therapeutically.

Transcranial Direct Stimulation of the Brain

Transcranial direct current stimulation (tDCS) of the brain has been reported as having psychoactive effects for at least 80 years. Modern reports were greeted with scepticism in many quarters. The latest studies seem to find some effectiveness vs. placebo in anxiety and depression for patients who sleep with one earlobe connected to the equivalent of a 12-volt battery and the other ear lobe connected to the other electrode of the battery. The clinical effects of tDCS have been accompanied by a large number of replicable findings on animal behaviour and neurophysiology. This is not a treatment of acute major depression and is still an ongoing area of investigation. The continuous stream of small positive clinical studies suggests that there really might be some effectiveness to this treatment although it is at this point difficult to define [12, 13].

The above four treatments are sometimes grouped together as "brain stimulation". It is not possible yet to know whether this collective terminology is justified by any common mechanism. Figure 9.1 summarizes the very different locations and current parameters of these "stimulation" treatments.

Clinical Vignettes

1. Arthur was a 56 year old man with no previous history of psychiatric disorder. Two months before referral he developed low mood, severe insomnia, loss of appetite with 15 pound weight loss, crying spells, and continuous suicidal

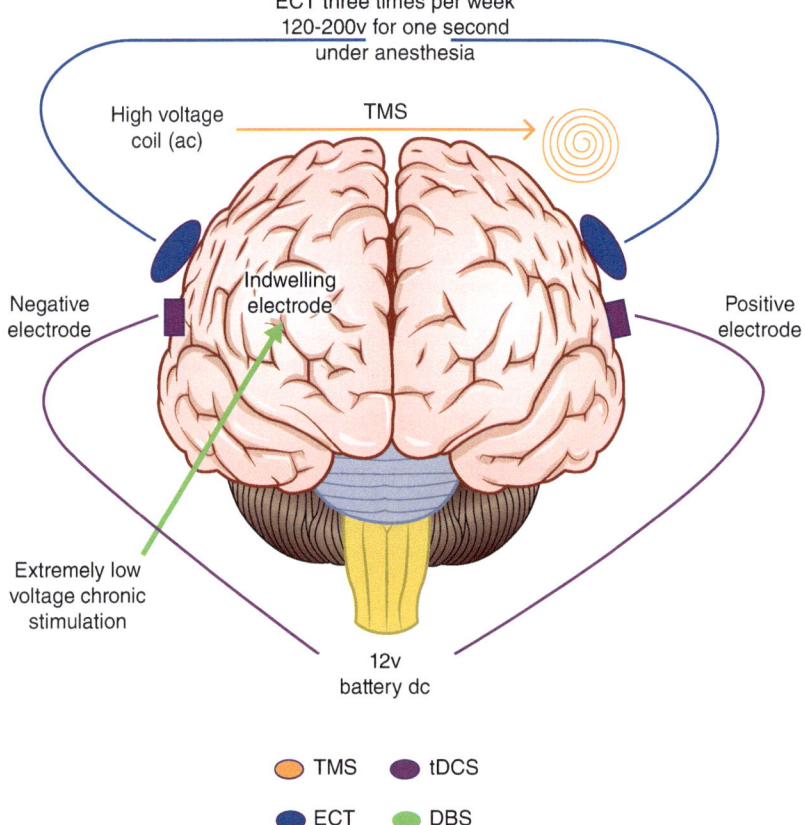

Fig. 9.1 Four different brain stimulation treatments

ideation and plans. There was no improvement with treatment with olanzapine 15 mg plus fluoxetine 40 mg daily for 4 weeks. He was not able to participate in psychotherapy and the family was not able to continue to provide 24 h daily supervision for suicide prevention. After one ECT treatment there was a lightening of mood and after two ECT treatments he was no longer suicidal. After 6 treatments he had regained weight and was sleeping. He was back to work at 1 month follow-up on the fluoxetine plus olanzapine at lower doses.

2. Edith was a 36 year old married woman who has been increasingly dysthymic in recent years and at some point slipped into diagnosable depression. She failed treatment with psychotherapy, with fluoxetine and with venlafaxine. She began daily TMS sessions on weekdays at a private clinic and slowly felt some relief and went back to work after 4 weeks of treatment. The dysthymia continued.

3. John was referred to a research study for deep brain electrode implantation at a tertiary hospital because of three suicide attempts and continuous unremitting depression. He was an adherent patient but had no relief of his depression.

4. Sarah was a fifty two year old married woman with a lifelong history of anxiety and depression and nothing seemed to help much. She bought a tDCS apparatus by mail order because of an internet advertisement. After sleeping nightly with this attachment as directed for a month, she feels much more energy, appetite, and sexual interest.

References

1. Sackeim HA, Prudic J, Devanand DP, Kiersky JE, Fitzsimons L, Moody BJ, et al. Effects of stimulus intensity and electrode placement on the efficacy and cognitive effects of electroconvulsive therapy. N Engl J Med. 1993;328(12):839–46.
2. Lisanby SH, Maddox JH, Prudic J, Devanand DP, Sackeim HA. The effects of electroconvulsive therapy on memory of autobiographical and public events. Arch Gen Psychiatry. 2000;57(6):581–90.
3. Squire LR, Zouzounis JA. ECT and memory: brief pulse versus sine wave. Am J Psychiatry. 1986;143(5):596–601.
4. Lerer B, Weiner R, Belmaker R. ECT: basic mechanisms. London: John Libbey; 1984.
5. Smith GE, Rasmussen KG Jr, Cullum CM, Felmlee-Devine MD, Petrides G, Rummans TA, et al. A randomized controlled trial comparing the memory effects of continuation electroconvulsive therapy versus continuation pharmacotherapy: results from the consortium for research in ECT (CORE) study. J Clin Psychiatry. 2010;71(2):185–93.
6. Belmaker RH, Fleischmann A. Transcranial magnetic stimulation: a potential new frontier in psychiatry. Biol Psychiatry. 1995;38(7):419–21.
7. Grisaru N, Yaroslavsky Y, Abarbanel J, Lamberg T, Belmaker R. Transcranial magnetic stimulation in depression and schizophrenia. Eur Neuropsychopharmacol. 1994;4:287–8.
8. George MS, Belmaker R. Transcranial magnetic stimulation in neuropsychiatry. Washington, DC: APA Press; 2000.
9. Lisanby SH, Morales O, Payne N, Kwon E, Fitzsimons L, Luber B, et al. New developments in electroconvulsive therapy and magnetic seizure therapy. CNS Spectr. 2003;8(7):529–36.
10. Fleischmann A, Prolov K, Abarbanel J, Belmaker RH. The effect of transcranial magnetic stimulation of rat brain on behavioral models of depression. Brain Res. 1995;699(1):130–2.

11. Lozano AM, Lipsman N, Bergman H, Brown P, Chabardes S, Chang JW, et al. Deep brain stimulation: current challenges and future directions. Nat Rev Neurol. 2019;15(3):148–60.
12. Chase HW, Boudewyn MA, Carter CS, Phillips ML. Transcranial direct current stimulation: a roadmap for research, from mechanism of action to clinical implementation. Mol Psychiatry. 2020;25(2):397–407.
13. Allenby C, Falcone M, Bernardo L, Wileyto EP, Rostain A, Ramsay JR, et al. Transcranial direct current brain stimulation decreases impulsivity in ADHD. Brain Stimul. 2018;11(5):974–81.

Chapter 10
Pain Medication and Opiate Addiction

Pain relief has been a universal goal of medicine since ancient times and almost every individual who experiences pain wants relief from this aversive sensation. Clearly psychological and social factors affect pain perception and also affect the subjective reporting of pain. Hypnosis, relaxation, distracting touch and other methods should always be considered in pain relief. Pharmacological relief of pain was discovered in ancient times in the form of extracts of the poppy plant which yielded opium, a powerful analgesic. Synthetic opium was first created in the 1930s as morphine. Since then, numerous similar compounds have been synthesized with increasing potency up to the present when substances like fentanyl are potent in microgram doses [1].

Among the discoverers of the opiate receptor was Solomon Snyder who used radioactively labelled opiate to identify the binding site. There are at least three types of receptors for opiates: mu, kappa and delta, each mediating slightly different functions. Using the binding site so identified, the endogenous ligands and neurotransmitters were discovered. The endogenous ligand for these receptors are enkephalins, or more broadly endorphins. The enkephalins are short peptides derived metabolically from the larger endorphins. The pain pathway ascends the spinal cord as the spinothalamic tract after receiving input from specialized pain sensing nerve endings distributed almost throughout the body. The spinothalamic tract ends in the thalamus which seems to be both the main center of pain perception and the pharmacologic anatomic location of analgesia by opiates at their receptors [2]. Figure 10.1 illustrates the spinothalamic ascending pain pathway. Of course the thalamus connects to many other areas of the brain and affects them, and much studied is the connection to dopamine-mediated reward systems. However, dopamine-blocking antipsychotic drugs do not block the analgesic or pleasure-giving effects of opiates in patients. Peripheral endorphins can be measured in blood and levels rise in certain circumstances such as exercise; they do not cross the blood-brain barrier; do not originate in the brain; and probably have no relationship to CNS analgesia or "high".

© The Author(s), under exclusive license to Springer Nature   123
Switzerland AG 2023
R. H. Belmaker, P. Lichtenberg, *Psychopharmacology Reconsidered*,
https://doi.org/10.1007/978-3-031-40371-2_10

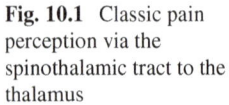
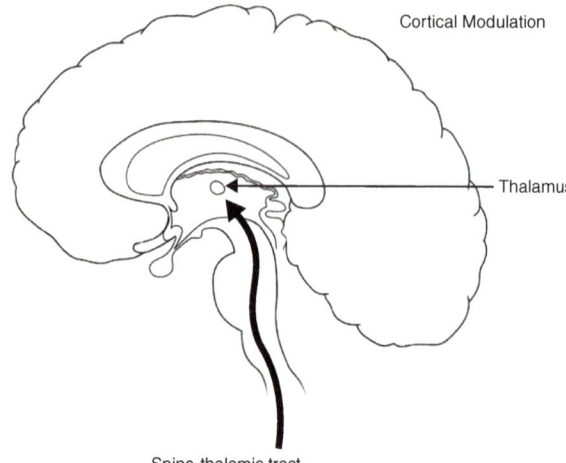

Fig. 10.1 Classic pain perception via the spinothalamic tract to the thalamus

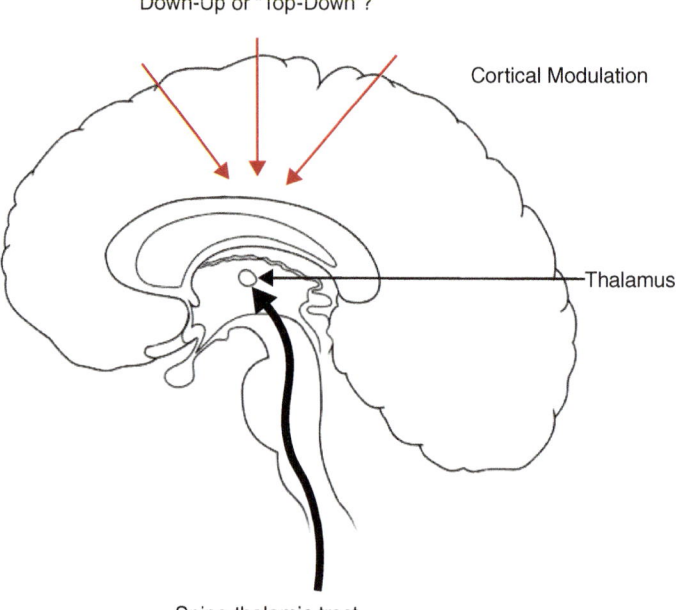

Fig. 10.2 Cortical top-down effects are critical in pain perception and behavior

Pain relief by opiates for acute pain such as bone trauma, myocardial infarction, metastatic cancer or pancreatitis is a consensus goal of all physicians. Dosing should be enough to relieve pain but the scheduling of doses is a specialized matter for each pain indication. Opiates always produce tolerance on chronic use and so the treatment plan for acute pain must always include the plan for reduction and eventual cessation of opiate pain relief. Opiates are almost never an appropriate treatment for chronic pain [3]. Figure 10.2 illustrates the important clinical fact that pain perception in the thalamus is highly modified "top-down" from the cortex and is affected by psychosocial inputs of many kinds.

The mechanism by which opiates produce tolerance has been studied extensively and involves many changes in the receptors but also in the post synaptic signal conduction systems. The tolerance seems to be a common phenomenon to all opiates and switching from one opiate to another does not alleviate tolerance. Tolerance develops over weeks to months and its mirror image is a withdrawal syndrome when the opiate is stopped in a tolerant individual taking high doses [4].

In addition to analgesia, opiates induce a pleasant sense of drowsiness and detachment from worry. Often individuals with acute pain alleviated by opiates can still locate and describe the pain but just don't care about it. For some individuals, but not all, and perhaps only for those who are genetically predisposed, the pleasant drowsiness and detachment achieved after opiate ingestion is a highly desirable goal in itself. These people may self-administer opiates after having been exposed to opiates for a therapeutic medical purpose or they might seek out opiates without ever having been treated with opiates medically. In rodent animal models, individual mice or monkeys will work hard to self-administer opiates. Often, they will do so in preference to food even if they are very hungry. On the other hand, some mice strains seem to be resistant to addiction to opiates. In humans it is difficult to estimate but perhaps 1/3 of humans are susceptible to opiate abuse if they are exposed whereas 2/3 might not even try the substance a second time if they were exposed to it once.

The withdrawal syndrome from an opiate tolerant state includes a generalized vague sense of pain in many areas of the body, especially the teeth and intestines, diarrhoea, and insomnia. The acute withdrawal syndrome is over in about a week, although of course this is dependent on the half-life of the particular opiate taken. Many studies in animals and some in humans suggest that certain elements of the addicted state at the cellular level remain latent in the brain for life and predispose to relapse to opiate addiction.

Most opiates are prescribed for accepted indications by non-psychiatrists. However, the world-wide problem of opiate addiction is treated mostly by psychiatrists [5].

Treatment of Opiate Addiction

The pharmacological treatment of opiate addiction is usually carried out in specialized centers that were once called "methadone clinics". Methadone is a long-acting opiate that can be administered orally on a daily basis under observation by a nurse; thus, avoiding the problem with prescription to home use that could be diverted by a treated patient to the black market and used for sale as opiates for abuse. Methadone can be used in the detoxification of an opiate addict since its long half life allows less frequent administration in an inpatient detoxification center. However, its unique role is in long term treatment as an opiate substitute. If given in high doses (around 100 mg per day) it often eliminates opiate cravings and rarely leads to further tolerance or need for dose escalation. Many long-term methadone patients return to a functional life using their daily methadone. The main side effects are constipation and erectile dysfunction, both of which are severe. The physician using methadone maintenance treatment should avoid the temptation to gradually lower the dose and "withdraw" the patient from methadone. When the dose goes below a certain level it will become possible for the patient to abuse opiates again and his daily methadone dose would then only serve as a convenient jumping off dose for him to use higher doses of other opiates [6].

Further developments in the long-term treatment of opiate abuse include buprenorphine, which is an opiate receptor agonist with antagonist effects at some opiate receptors at some doses. Buprenorphine therefore is less likely to be diverted into the black market but on the other hand patients find that it leaves them with irritability and dysphoria due to lack of adequate stimulation at opiate receptors. Therefore, compliance may be higher in some groups with methadone maintenance than with buprenorphine maintenance. In some settings buprenorphine is combined in the same tablet with naloxone, an opiate antagonist (see below), which further reduces risks of diversion into the black market.

Methadone or buprenorphine maintenance must always be combined with psychosocial treatment including avoidance of situations where the patient had previously been conditioned to use opiates. Former opiate addicts need employment rehabilitation and treatment of any illness acquired during opiate addiction, including hepatitis and AIDS. Psychiatrists must be closely involved with this treatment as the goals of rehabilitation are different for each patient and his family. The psychiatrist must himself be able to find professional satisfaction in a treatment that is symptomatic and that does not involve total elimination of the methadone as an opiate in the patient's life. Some social criticism by missionary groups who condemn methadone maintenance as continued addiction must be endured. The well-trained psychiatrist usually realizes that realistic goals that are attainable are more consistent with medicine's raison d'etre than unattainable goals [7].

Side Effects

The side effects of all opiates as mentioned above include gastrointestinal constipation and erectile dysfunction. The constipation can usually be managed with bulk fiber anti-constipation treatments. The erectile dysfunction may sometimes respond to sildenafil. The most serious side effect of all of opiates derives from their capacity to cause respiratory depression in high doses. This sometimes leads to an acceptable and painless death in terminal cancer patients for whom relief of untreatable pain is an absolute necessity and specialist physicians who deal with opiates in hospice care often consult with psychiatrists about their feelings. More problematic is the respiratory depression induced by an opiate that the drug abuser assumed was a diluted dose but, which unbeknownst to him, was a higher dose [8]. These accidental overdoses are responsible for a huge number of deaths in the United States today across all social classes, ages and sex categories. The respiratory depression caused by opiates is easily reversed by an acute intravenous dose of an opiate antagonist such as naloxone [9]. Naloxone must be available to all emergency medical staff who might be called to an overdose incident and perhaps these antagonists should be available in an easily administered method such as intranasal in all places where opiate addicts congregate and do self-injection of opiate drugs [10]. Since the antagonist naloxone has a relatively short half-life, patients who are revived may sink back into an opiate coma with respiratory depression a few hours later. Follow up and repeat naloxone is absolutely necessary [11].

Causes of Opiate Addiction

Opiate addiction has existed since the discovery of the properties of extracts of the poppy plant. Historically opiate addiction in China in the 1800s was encouraged by the British for financial reasons and led to the Boxer rebellion in an attempt by the Chinese to control their own national opiate consumption. Another wave of opiate addiction occurred in the USA in the late 1800s among young mothers when opiates were given for obstetric pain in an unregulated and overly free manner by doctors, particularly to upper socioeconomic class childbearing women. At some points in American history opiate addiction was associated with Afro-American race and low socioeconomic status. The most recent wave of opiate addiction cuts across socioeconomic lines and seems to be, at least partly, a product of the development of oxycodone and commercial advertisement and aggressive marketing by Purdue Pharmaceutical company [12]. This campaign implied that oxycodone was somehow not an opiate and not addictive [13]. Coincidently or not, it came simultaneously with an academic reawakening of interest in the problem of pain in general practice. Many academic articles were published claiming that pain was being ignored by family physicians. Some medical insurance companies began to require that doctors declare that they have asked every patient whether he has any pain and treated any such pain adequately in order to receive reimbursement for the patient's visit [14]. These factors combined to cause an explosion both in use of oxycodone and in opiate related deaths as patients who became addicted used the drug recreationally and sometimes purchased doses in the black market that were overdoses and respiratory depressants. This pandemic of opiate abuse is not yet over and a clear solution at the public health level is not yet in sight [15]. The answer is clearly not the ban of all opiate use because there are still no alternatives to opiates in severe acute pain. However, use of opiates for chronic pain, such as lower back pain, fibromyalgia, chronic recurring tension headaches and other such disorders should be strongly avoided because they present a great risk of addiction. Non-steroidal anti-inflammatory drugs (NSAIDS) are inhibitors of the cyclo-oxygenase enzyme and offer some analgesia along with anti-inflammatory effects. They are first line treatment for mild acute pain or chronic pain along with psychosocial treatments and before opiates of course [16]. Cannabis is not effective in replacing opiates, neither in opiate-addicted individuals nor in severe pain requiring opiates (see Chap. 16 on cannabis).

Clinical Vignettes

1. Jacob was a 45 year old truck driver who was first exposed to opiates for his back pain by a prescribing physician. He loved the feeling it induced and within months was taking three times the prescribed dose, obtaining the extra via the black market. He crashed his truck. A psychiatrist was asked to evaluate whether he was responsible for his actions legally at the time of the crash.
2. Sarah was 23 when she was offered to share the contents of a syringe by her boyfriend, an addict who used opiates regularly. She stopped breathing after the intravenous injection and died.
3. Netaniel was a 57 year old terminal pancreatic cancer patient in hospice. Morphine was necessary at high doses. A new nurse asked that a psychiatrist be called, telling him that she was afraid the patient would become addicted.
4. Ariel was 32 and in methadone maintenance after two jail terms for selling opiates. His dose was 80 mg orally under direct observation. He asked the psychiatrist for a dose lowering plan.

References

1. Babu KM, Brent J, Juurlink DN. Prevention of opioid overdose. N Engl J Med. 2019;380(23):2246–55.
2. Haight BR, Learned SM, Laffont CM, Fudala PJ, Zhao Y, Garofalo AS, et al. Efficacy and safety of a monthly buprenorphine depot injection for opioid use disorder: a multicentre, randomised, double-blind, placebo-controlled, phase 3 trial. Lancet. 2019;393(10173):778–90.
3. McQueen K, Murphy-Oikonen J. Neonatal abstinence syndrome. N Engl J Med. 2016;375(25):2468–79.
4. Fairley M, Humphreys K, Joyce VR, Bounthavong M, Trafton J, Combs A, et al. Cost-effectiveness of treatments for opioid use disorder. JAMA Psychiatry. 2021;78(7):767–77.
5. Compton WM, Jones CM, Baldwin GT. Relationship between nonmedical prescription-opioid use and heroin use. N Engl J Med. 2016;374(2):154–63.
6. Schuckit MA. Treatment of opioid-use disorders. N Engl J Med. 2016;375(4):357–68.
7. Kalkman GA, Kramers C, van Dongen RT, van den Brink W, Schellekens A. Trends in use and misuse of opioids in The Netherlands: a retrospective, multi-source database study. Lancet Public Health. 2019;4(10):e498–505.
8. D'Onofrio G, O'Connor PG, Pantalon MV, Chawarski MC, Busch SH, Owens PH, et al. Emergency department-initiated buprenorphine/naloxone treatment for opioid dependence: a randomized clinical trial. JAMA. 2015;313(16):1636–44.
9. Lott DC. Extended-release naltrexone: good but not a panacea. Lancet. 2018;391(10118):283–4.
10. Ortega R, Nozari A, Baker W, Surani S, Edwards M. Intranasal naloxone administration. N Engl J Med. 2021;384(12):e44.
11. Lee JD, Nunes EV Jr, Novo P, Bachrach K, Bailey GL, Bhatt S, et al. Comparative effectiveness of extended-release naltrexone versus buprenorphine-naloxone for opioid relapse prevention (X:BOT): a multicentre, open-label, randomised controlled trial. Lancet. 2018;391(10118):309–18.
12. Podolsky SH, Herzberg D, Greene JA. Preying on prescribers (and their patients) - pharmaceutical marketing, iatrogenic epidemics, and the Sackler legacy. N Engl J Med. 2019;380(19):1785–7.

13. Larance B, Dobbins T, Peacock A, Ali R, Bruno R, Lintzeris N, et al. The effect of a potentially tamper-resistant oxycodone formulation on opioid use and harm: main findings of the National Opioid Medications Abuse Deterrence (NOMAD) study. Lancet Psychiatry. 2018;5(2):155–66.
14. Davis CS. The Purdue pharma opioid settlement - accountability, or just the cost of doing business? N Engl J Med. 2021;384(2):97–9.
15. Cicero TJ, Ellis MS, Surratt HL. Effect of abuse-deterrent formulation of OxyContin. N Engl J Med. 2012;367(2):187–9.
16. Baker DW. History of the joint Commission's pain standards: lessons for Today's prescription opioid epidemic. JAMA. 2017;317(11):1117–8.

Chapter 11
Stimulant Drugs: Are They Specific for Attention Deficit Disorder or Are They Abused and Overused?

The first stimulants amphetamine and cocaine were synthesized in the late nineteenth century and found to have powerful effects in humans to increase self-confidence, aggressivity, decrease need for sleep, increase vigilance and increase sense of self confidence [1]. Cocaine was famously used by Freud and was praised by Freud for its pleasure giving ability before he realized its addictive potential and gave it up. The addictive potential of these compounds was a theme of much literature throughout the twentieth century and the chronic use of amphetamine or methylphenidate was known to cause a paranoid psychosis with delusions of persecution and ideas of grandeur not unlike those in schizophrenia [2]. Parallel to the literature on the stimulant and addictive and psychosis inducing properties of these compounds in adults, a parallel and seemingly disconnected universe of clinical studies beginning in the 1930s reported that stimulant compounds like amphetamine and methylphenidate could calm brain damaged patients with hyperkinetic syndromes. These syndromes seemed to paradoxically be quieted by stimulant drugs. Shortly thereafter Bradley in the 1930s found that hyperactive children were quieted by amphetamines. The children he described were organically impaired and their hyperkinesia was presumed to be due to deficits or damage in the basal ganglia. Later Wender [3] described the concept of "minimal" brain dysfunction [4] which consisted of a syndrome of soft neurological signs along with hyperactivity and attention deficits. The soft neurological signs included abnormalities such as dysdiadokinesia, minor tremors, and asymmetric reflexes that are normal at particular ages but represented delayed neurodevelopment if still present at ages such as 7, 8 or 9. Wender proposed that this minimal brain dysfunction syndrome marked an amphetamine responsive illness. Wender and others reconciled the euphorigenic, addictive, and psychotogenic properties of stimulants in adults with their apparent therapeutic properties to reduce hyperactivity in minimal brain disorder children by postulating that these compounds have "paradoxical" effects in children that disappear in adolescence. They at first warned against the use of stimulant drugs after adolescent changes in brain development.

R. H. Belmaker, P. Lichtenberg, *Psychopharmacology Reconsidered*, https://doi.org/10.1007/978-3-031-40371-2_11

The concept that stimulant drugs might have a paradoxical calming effect in the prepubertal child has been studied heavily over the years [5]. It has been difficult to prove or disprove. Others have claimed that the calming effect of stimulant drugs is pathognomonic to minimal brain dysfunction but this concept has also been difficult to prove or disprove over the years. It remains in all honesty a powerful paradox that the same drugs can be described as euphorigenic, stimulating, aggression-inducing, highly addictive and psychotogenic in many clinical contexts and on the other hand calming, therapeutic and inducing enhanced concentration, focus and learning ability in other contexts. See Fig. 11.1. There are many papers on amphetamines, methylphenidate and cocaine addiction and psychosis that do not have a single reference to the literature on their therapeutic potential. Conversely, there are numerous papers on the use of amphetamine, methylphenidate and their derivatives in childhood hyperactivity that do not mention the literature on the drugs' euphorigenic, addictive and psychotogenic dangers. This chapter will attempt to discuss the use of stimulant drugs in ADHD without ever losing the perspective and awareness of these drugs' euphorigenic, addictive and psychotogenic potential. As a spoiler, we must state already at this point that we do not solve the existence of this paradox by the end of this chapter in any way. In animal models, injection of amphetamine, methylphenidate or cocaine have similar effects to increase activity of rodents, to increase aggressivity, to decrease need for sleep, to decrease need for food and to increase sexual activity. As is well known, prepubertal rodents like prepubertal humans and all other mammals, are much more active during their awake hours than their adult counterparts See Fig. 11.2. There is some evidence that amphetamine and similar drugs have a paradoxical effect in prepubertal animals to reduce activity rather than increase it but this is dependent on conditions and dose [6].

On a biochemical level amphetamine and methylphenidate directly release noradrenalin, serotonin and dopamine from the presynaptic terminal, effectively pushing the vesicles of these compounds to fuse with the presynaptic membrane and

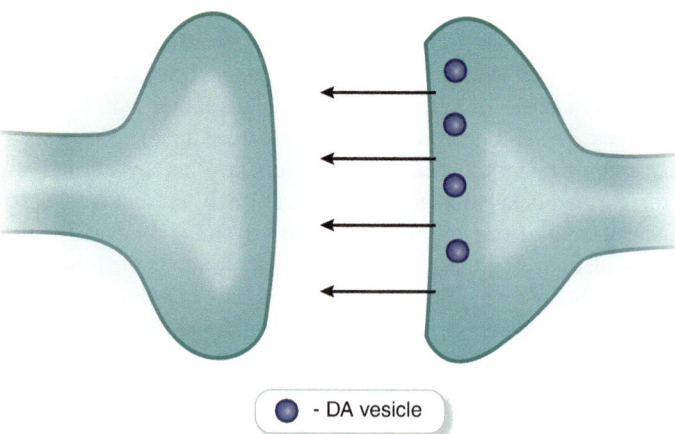

● - DA vesicle

Fig. 11.1 Stimulants cause release of monoamines from the presynaptic neuron vesicles

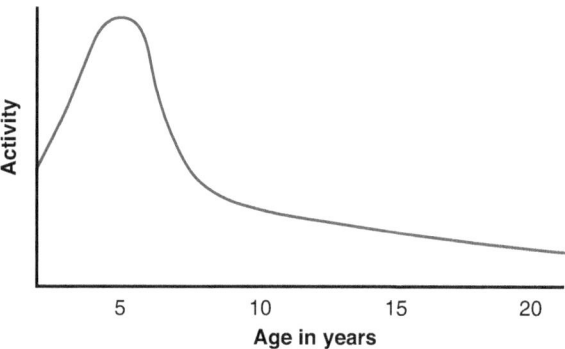

Fig. 11.2 Average normal activity is highest in childhood

empty their contents into the synapse. Some compounds affect release of noradrenaline more than serotonin but all release dopamine. Their effects can be entirely blocked in most models by blocking the post synaptic receptor with an antagonist such as haloperidol before the amphetamine like stimulant is administered. Belmaker [7] found that haloperidol intravenously can totally block the euphorigenic effects of amphetamine in humans thus suggesting that dopamine is the key actor in the stimulant actions. Cocaine, in addition to directly releasing noradrenalin, serotonin and dopamine into the synapse, seems to have the additional imipramine-like action of preventing the re-uptake of these compounds, thus further enhancing their effective responses in the synapse and action at the post synaptic receptors. However, cocaine effects are blockable by haloperidol or other dopamine receptor blockers in most animal models and human models as well [8].

The above discussion is a difficult and unusual introduction for a chapter on the therapeutic use of stimulants. Indeed, as one looks back on this history one is amazed. After Wender's description of MBD (minimal brain dysfunction) several other child psychiatry groups enhanced and expanded his concept and disconnected it entirely from any concept or need for pre-existing brain damage, minimal or otherwise. These groups created the concept of childhood hyperactivity disorder and later linked it to attention deficit disorder, finally resulting in a concept of attention deficit disorder with or without hyperactivity. This new diagnosis exploded in reported prevalence and reports in recent years using liberal definitions have found 5–15% of children to have such a symptom cluster, more in boys than in girls. The syndrome provided an indication for use of stimulant treatment which occurs in 5–10% of children in some communities. The use of amphetamine or other stimulant treatment for a diagnosis of ADHD is associated in complex ways with social class. On the one hand it has been reported that children from lower socioeconomic classes in the US are more likely to be diagnosed with ADHD on complaint of a teacher and to be prescribed a stimulant. On the other hand, affluent communities often have the highest percentage of children receiving stimulant treatment, sometimes up to a third of elementary school children receiving stimulant treatment. Stimulant treatment in some communities has become a sign of parental concern about their children's competitiveness in school and some drug advertisements have suggested to parents that their children will receive better grades if treated with stimulants. Numerous stimulants are currently available in clinical practice. Amphetamine is available both in the racemic mixture of mixed salts; as the pure d-amphetamine, as lisdexamfetamine, a prodrug that is converted into amphetamine by an enzyme in the red blood cell. Methylphenidate (Ritalin) is available as originally discovered as a compound whose pharmacodynamics lasts for 4 or 5 h or in slow release form Concerta or as a different slow release form Ritalin SR. The large number of these formulations has paralleled their rapidly increasing use in clinical child psychiatry.

The Specificity of Stimulants in ADHD

This textbook is not the appropriate place to detail the creation of the concept of ADHD by the DSM committee and the evidence or lack of it for the validity and internal consistency of this spectrum of symptoms. Motor activity level exists on a normal continuum in the population of children. See Fig. 11.3. Some children are more active than others and a small number of children are extremely active.

There is no clear cut off or bimodal distribution beyond which a child can be clearly diagnosed as abnormally hyperactive. The abnormality of a child's hyperactivity is a relative function of two factors: his activity level and the demands of his environment. A child in a strict school environment that demands his attention for eight full hours of lessons is more likely to be diagnosed as hyperactive than a child with a similar activity level but who studies in a more open school environment. Similarly, the level of concentration ability of children varies along a normal curve with no cut-off or bimodal distribution beyond which a child can be clearly defined as having an attention deficit disorder.

As with activity level the existence of an attention deficit disorder requires a co-occurrence of two factors: A low attention level in a particular child and a high demand for attention in a competitive academic school with academically

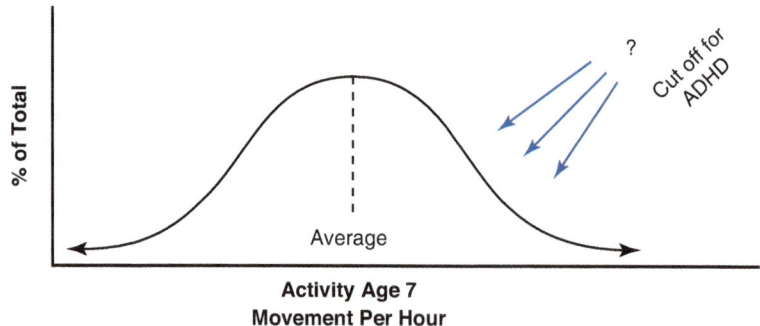

Fig. 11.3 Activity levels in childhood vary around a normal curve, with no clear cut-off point

competitive and demanding parents or social group. The relationship of attention deficit disorder to hyperactive disorder is only very partial and their consolidation into a single diagnosis is probably arbitrary according to current evidence. No specific gene has been found for these two traits although they both have significant heritability as does almost all of normal human variation in physical and behavioral traits. No specific or diagnostic finding on brain MRI or brain biochemistry has been found for children with ADHD. Many reports have found statistically significant differences in the size of various areas of the brain such as the basal ganglia, the hippocampus or the frontal lobes. Most of the findings are not replicable. Some that are replicable are clinically insignificant because the overlap in the patient population and the control population is so great that the finding has no diagnostic use. Common validating tests for ADHD such as the computerized continuous performance test generate quantitative data and cut-offs are given by several companies and are used in some jurisdictions as justification for stimulant treatment or special school help. No follow up data validates the use of such tests as diagnostic therapeutic thresholds for stimulant treatment.

Effectiveness of Stimulant Treatment in ADHD

Amphetamine or methylphenidate treatment of hyperactive children reduces their activity level in a clinically significant degree. In some circumstances this can be a great relief for the caretaker or teacher. Sometimes it can allow the child to remain in the school of his choice rather than be expelled. In that sense stimulant treatment of childhood hyperactivity can be effective and justifiable. Stimulant treatment of poor learning in children is less clearly efficacious. Specific learning disabilities such as dyslexia have not been proven to be helped by stimulants at all although stimulant use was once a great hope for these troublesome disorders. Some children with daydreaming and inability to concentrate in class may be helped by a stimulant which allows them to focus more effectively. However, this effect is less pronounced than the dramatic effect on the hyperactive child. The effect to increase focus in some daydreaming children is easier to conceptualize in the context of the classic effect of stimulants to increase activity, persistence, energy and focus. The effect of stimulants to decrease activity in the hyperactive child is more difficult to conceptualize although some have speculated that hyperactivity is essentially a lack of focus and that stimulant treatments reduce activity by allowing an increased focus on the central issues of the environment, usually the teacher.

Side Effects of Stimulant Drug Treatment

Treatment with amphetamine, methylphenidate, and similar drugs can cause tics de novo or exacerbate existing tic syndrome or Tourette's syndrome. Blood pressure elevation is possible though uncommon but anorexia and some growth retardation are frequent. The weight and height are usually regained during a summer drug holiday.

Psychological Side Effects of Stimulant Treatment

In classic adult studies of stimulants, healthy volunteers given d-amphetamine at doses of 10 mg every few hours became sleepless by day one and showed signs of psychotic thinking by day three to five after total dosages between 100 and 700 mg of d-amphetamine, not too far from clinical dosages [9]. Children receiving daily doses of up to 60 mg of amphetamine salts rarely become psychotic but it is a danger that should not be discounted [10]. A much more frequent side effect is sadness, perhaps related to the fact that these monoamine releasing drugs do over time deplete monoamines in the presynaptic cell. It is well known that amphetamine abusing adults and cocaine abusing adults crash into sadness when they stop but also sometimes become sad while they continue to use the compounds and are driven to use increasingly large doses therefore [11].

Children using stimulants to control hyperactivity may become over focused and unable to switch from school work to hobbies and play. They may become "quenched" or look slightly unresponsive. Parents and teachers should be aware of this side effect which is a high price for a child to pay. It is rarely improved by merely switching to another stimulant with the same mechanism of action and more often is an indication for stopping stimulant treatment.

While addiction is rare in children whose dose and administration are controlled by parents, addiction can occur in adolescents continuing on amphetamine or other stimulant treatment. While the concept that puberty induces a change in reactivity to stimulants such that they are no longer paradoxically "calming" has not been proven, many clinicians do see that adolescents become dependent on stimulant drugs, need it to get moving in the morning and sad without their drugs. The clinical report begins to be more typical of a cocaine or amphetamine addicted adult than a stimulant treated prepubertal hyperactive child. Adolescents, even those who benefited from stimulant treatment of hyperactivity in childhood, should be weaned off of stimulant drug treatment because its dangers are greater than its possible benefits in most cases.

Comorbidity

ADHD is highly comorbid with oppositional defiant disorder (ODD), specific learning disorder (dyslexias), anxiety disorders, conduct disorder and intellectual disability. Stimulant treatment is not effective in these comorbidities. A common clinical mistake is to treat a disobedient child with stimulants when the parents are mainly concerned with the disobedience, even if some ADHD is indeed present. Everyone will be disappointed.

Use of Stimulants for Learning Enhancement

Stimulants do not change measured IQ in children or adults. Many individuals are disappointed with their own realistic prospects of academic achievement and amphetamine treatment does not alter those prospects and should not be proposed as such. Many college students use stimulant drugs in the hope of better study results, longer study concentration hours, or better memory retention. While some laboratory evidence supports the idea that stimulants can have such effects, the effects are clinically small and of no real value in educational success. More likely, students are being attracted by the sense of energy, self-worth, and optimistic self-evaluation induced by the euphoriant effect of stimulants. This is self-deception and a true cultural danger. It is sometimes truly disturbing to hear a student say: "I love amphetamine so much, that proves I have ADHD". Stimulants do not treat the depression that sometimes follows academic failure, and should not be offered as a sure cure to reverse the failure on a second try.

Long Term Outcomes

Few long-term follow up studies of children treated with stimulant drugs vs. controls exist. Those that do exist do not find that children treated with stimulant drugs have an increased incidence of drug abuse as adults, a finding that has been reassuring to many. However, it should be noted that such studies are restricted to carefully chosen children who were given amphetamines for strictly defined criteria under strict research standards. It can be questioned whether the large numbers of children who are given amphetamines for more vague diagnoses and with less strict follow up in today's conditions will have an equally danger free outcome [12].

Non Monoamine-Releasing Stimulants

Interestingly, some drugs that do not directly release monoamines are stimulants and have some efficacy in childhood hyperactivity or attention deficit. One is modafinil [13], an awakeness-promoting compound of unclear mechanism. Another is atomoxetine, a simple imipramine-like reuptake inhibitor that has been channeled into a marketing niche as an amphetamine alternative for ADHD. Lastly, caffeine, usually reported to act via adenosine receptors in the brain, is mankind's most commonly used stimulant. All these compounds have some efficacy in reducing hyperactivity in children but are generally accepted as less powerful than amphetamine or methylphenidate. The fact that various stimulants with very diverse neurochemical mechanisms are helpful in hyperactivity of childhood may suggest that neurochemistry is not the correct level of explanation for these effects: maybe CNS stimulation can have many mechanisms and a very nonspecific benefit in ADHD.

Use of Stimulants in ADHD Diagnosed Adults

Clinicians studying alcohol abuse epidemiology have long noted the importance of an "alcohol culture" where drunkenness is acceptable and alcohol is a necessary part of life as keys to a society with excessive use. It was long a gossip item that famous actors and actresses used heavy sedatives to sleep at night and amphetamines to get moving in the mornings. But in recent years one hears this at parties and in family gatherings as a frequent story, with no need to refer to a medical "disease" of ADHD. Stimulant use has been normalized in Western culture [14]. The diagnosis of ADHD in adults represents a 180 degree turnaround from the historical concept of amphetamine as having a paradoxical calming reaction in hyperactive children, in contrast to its known euphorigenic and activating effects in adults. Amphetamine and methylphenidate treatment are actively sought out by many patients who come to psychiatrists, neurologists and family physicians. Often the individual has tried stimulants on his/her own and likes it and wants it. The amount of amphetamine and methylphenidate prescribed by physicians is a fraction of that consumed by the public, without even considering the use of cocaine (illegal) or methamphetamine (illegal). It makes sense that some adults who had ADHD as children have lost the hyperactivity at puberty but remain with attention deficit [15]. It is a difficult decision for the psychiatrist and for the profession to separate this group of people, if they can be defined, from individuals who are seeking amphetamines to boost confidence, feel euphoric, feel sexually potent and strong and are at high risk of addiction.

Pregnancy and Lactation

While the catastrophic risks once predicted for "cocaine babies" were exaggerated, stimulants are not safe in pregnancy or lactation. While it is difficult to separate studies that found malformations with doses of amphetamine, methylphenidate or methamphetamine in abuse with results for lower dose treatment of adult ADHD, safety requirements of the level required for safeguarding an unborn human being are clearly not met. Moreover, ADHD treatment targets are for diseases that are not life-threatening and drug holidays are often advised clinically for other reasons. Pregnant or lactating mothers should not take stimulants [16].

How Can Dopamine Enhancement Help with Hyperactivity and Dopamine Reduction Improve Psychosis?

Some have speculated and presented indirect evidence to the effect that the therapeutic dopamine blockade in the treatment of psychosis occurs in different areas of the brain than the therapeutic dopamine enhancement that occurs in stimulant treatment of childhood hyperactivity. However, both stimulants and antipsychotics reach all areas of the brain and both work on all dopamine synapses. Antipsychotic dopamine blockade does not cause hyperactivity. Low dose clinical stimulant treatment of hyperactivity does not regularly cause psychosis. Most surprisingly, risperidone a powerful dopamine D-2 receptor blocker and antipsychotic, is widely prescribed for conduct disorder and ADHD in children often together with dopamine releasing stimulants [17]! This is hard to understand in any theoretical framework. Perhaps it is possible to see these theories like quantum mechanics and Einsteinian relativity in physics, two completely contradictory theories that are both true but in different realms, one in subatomic particles and the other in galactic motion. A well-known Professor of Psychiatry at Duke University, John Rhoads, used to say: No theory is perfect: Use each one where it works.

Paradoxes
- Why should stimulants reduce activity?
- Do children with ADHD get addicted?
- Why do ADHD children on stimulants get sad?
- Why are there differences in usage from area to area?

Vignettes

1. Richard, a third grade pupil, had been highly active since age two but his parents preferred to avoid pharmacological treatment. He needed special lessons to acquire reading because of moderate dyslexia and was still wetting the bed in third grade. In third grade his activity level began disturbing classroom studies as he got up several times during each lesson and needed to walk around the room. He responded to low dose Ritalin therapy of 10 mg each morning and noon with reduced activity but no changes in his other symptoms.

2. Jeffery was the first grade student of two busy professional parents who found him impossible to control by age four because he interrupted all of their well-planned evening and weekend activities as well as being a nuisance at school. He was a know-it-all in his responses to teachers, misbehaved when sent to the principal, teased other students and used basketballs in student intermission time to break windows. His parents began stimulant treatment in preschool with Ritalin which caused sadness, continued to lisdexamfetamine (Vyvanse) which caused a worrisome tic syndrome and then onto mixed amphetamine salts which caused anorexia and sleeplessness. After almost 2 years of multiple drug trials looking for the perfect drug to fix their kid, they reached atomoxetine which had no side effects and seemed to leave the child in a state very similar to that in which he had first been sent for treatment. The parents were offered parental counseling but said they were too busy.

3. George is a 37 year old man with poor job performance lifelong who feels great when he takes amphetamine. It fills him with confidence and he says that it helps his concentration. He has been diagnosed as adult-onset ADHD. His dose seemed to continually increase as he demanded the substance from several physicians. In recent months he attributes his poor job performance to people at work who have plotted against him because of his good looks and many girlfriends.

References

1. Posner J, Polanczyk GV, Sonuga-Barke E. Attention-deficit hyperactivity disorder. Lancet. 2020;395(10222):450–62.
2. Ashok AH, Mizuno Y, Volkow ND, Howes OD. Association of Stimulant use with Dopaminergic Alterations in users of cocaine, amphetamine, or methamphetamine: a systematic review and meta-analysis. JAMA Psychiatry. 2017;74(5):511–9.
3. Wender PH. Minimal brain dysfunction in children. Diagnosis and management. Pediatr Clin N Am. 1973;20(1):187–202.
4. Eisenberg J, Brecher-Fride E, Weizman R, Ebstein RP, Belmaker RH. Dopamine receptors in a rat model of minimal brain dysfunction. Neuropsychobiology. 1982;8(3):151–5.
5. Chan E, Fogler JM, Hammerness PG. Treatment of attention-deficit/hyperactivity disorder in adolescents: a systematic review. JAMA. 2016;315(18):1997–2008.

6. Ching C, Eslick GD, Poulton AS. Evaluation of methylphenidate safety and maximum-dose titration rationale in attention-deficit/hyperactivity disorder: a meta-analysis. JAMA Pediatr. 2019;173(7):630–9.

7. Wald D, Ebstein RP, Belmaker RH. Haloperidol and lithium blocking of the mood response to intravenous methylphenidate. Psychopharmacology. 1978;57(1):83–7.

8. Cortese S. Pharmacologic treatment of attention deficit-hyperactivity disorder. N Engl J Med. 2020;383(11):1050–6.

9. Griffith J, Cavanaugh J, Held J, Oates J. Dextroamphetamine. Arch Gen Psychiatry. 1972;26:97–100.

10. Moran LV, Ongur D, Hsu J, Castro VM, Perlis RH, Schneeweiss S. Psychosis with methylphenidate or amphetamine in patients with ADHD. N Engl J Med. 2019;380(12):1128–38.

11. Nourredine M, Gering A, Fourneret P, Rolland B, Falissard B, Cucherat M, et al. Association of Attention-Deficit/hyperactivity disorder in childhood and adolescence with the risk of subsequent psychotic disorder: a systematic review and meta-analysis. JAMA Psychiatry. 2021;78(5):519–29.

12. Cortese S. Psychosis during attention deficit-hyperactivity disorder treatment with stimulants. N Engl J Med. 2019;380(12):1178–80.

13. Belmaker RH. Modafinil add-on in the treatment of bipolar depression. Am J Psychiatry. 2007;164(8):1143–5.

14. Kazda L, Bell K, Thomas R, McGeechan K, Sims R, Barratt A. Overdiagnosis of attention-deficit/hyperactivity disorder in children and adolescents: a systematic scoping review. JAMA Netw Open. 2021;4(4):e215335.

15. Cortese S, Adamo N, Del Giovane C, Mohr-Jensen C, Hayes AJ, Carucci S, et al. Comparative efficacy and tolerability of medications for attention-deficit hyperactivity disorder in children, adolescents, and adults: a systematic review and network meta-analysis. Lancet Psychiatry. 2018;5(9):727–38.

16. Huybrechts KF, Bröms G, Christensen LB, Einarsdóttir K, Engeland A, Furu K, et al. Association between methylphenidate and amphetamine use in pregnancy and risk of congenital malformations: a cohort study from the international pregnancy safety study consortium. JAMA Psychiatry. 2018;75(2):167–75.

17. Weizman A, Weitz R, Szekely GA, Tyano S, Belmaker RH. Combination of neuroleptic and stimulant treatment in attention deficit disorder with hyperactivity. J Am Acad Child Psychiatry. 1984;23(3):295–8.

Chapter 12
Childhood Psychopharmacology: Modern Parent's Salvation or Danger to the Brains of the Future Generation?

The most common pharmacologically treated disorder in childhood is hyperactivity and attention disorder; this has been covered in Chap. 11. Another specific much less common but typical neurodevelopmental syndrome, Tourette's, is covered in Chap. 19. This chapter covers several syndromes that illustrate the limits of psychopharmacology for mild as well as severe syndromes.

Enuresis

Enuresis is a syndrome of night-time urinary incontinence after the age when children usually achieve bladder control at night [1]. At age three all children are bed wetters; by first grade about 25% remain bed wetters and at age 18 about 2% are bedwetters. Children who achieve night-time bladder control may revert to bed wetting in winter. The development of bladder control is a neurodevelopmental process that occurs at different rates in different children: Epidemiological rates can be fitted on an exponential curve such that every 3 years about half of children cease wetting the bed. It is more frequent in boys than girls. There is a large genetic component but psychosocial factors are also important. Avoidance of evening drinking, waking up the child for the toilet before the parent's bedtime, and clear parental expectations are clearly important. Some parents view this as merely a laundry problem rather than a medical problem. Certainly, invasive procedures such as IVP (intravenous pyelogram) or cysto-urethrograms are unnecessary unless the pediatrician has noticed particular warning signs of infection or urethral malformation. A bell and pad conditioning treatment is often effective after the age of five but will not be discussed here. Psychopharmacological approaches began with the use of imipramine which many studies find to be effective in reducing childhood enuresis. If used, it should be stopped for a couple of weeks every few months to check if the enuresis has remitted on its own. It was speculated that the effect of imipramine

R. H. Belmaker, P. Lichtenberg, *Psychopharmacology Reconsidered*, https://doi.org/10.1007/978-3-031-40371-2_12

might merely be an anticholinergic, atropine-like effect on the bladder sphincter but controlled studies ruled this out.

Another psychopharmacological treatment of enuresis uses a long-acting analogue of the natural antidiuretic hormone (ADH). The long-acting synthetic peptide (DDAVP, desmopressin) is usually administered by nasal spray or drops before bedtime where it is absorbed in the upper regions of the nasal epithelium. Oral forms, especially effective as sublingual oral dispersible tablets, are now available despite earlier concerns that the peptide would not be absorbed from the GI tract. DDAVP reaches the kidney via the blood stream and reduces kidney fluid excretion for several hours and can be very effective in preventing enuresis. Some parents are concerned about taking "hormones" but even though the ADH peptide is also a CNS neurotransmitter, evidence suggests that exogenously administered DDAVP does not reach the brain. Psychosocial consequences of bed wetting can be severe in some cultural groups and this can be an indication for treatment psychopharmacologicaly for limited periods of time until night-time continence is achieved [2].

Disruptive Mood Dysregulation Disorder

Temper outbursts are common in 3 year olds as every parent knows. As the child becomes older, temper outbursts can be more difficult to control, frightening to parents and siblings and a reason for failure at school. The incidence of children of school age with irritability and rage attacks has seemed to increase rapidly in North America over recent decades. It is difficult to know if this reflects decreased parental authority, smaller families with resultant parental overconcern about each child, changing cultural messages that suggest that children have human rights so broad that any frustration is unacceptable, or increased computer and television exposure to violence and rage. Some prominent researchers presented evidence that irritability and rage attacks in children and adolescents constitute childhood bipolar disorder, even if discrete periods of mood elevation and depression were not present. This led to the trial of mood stabilizers including lithium, valproate, carbamazepine and particularly second generation antipsychotics in these children. While lithium, valproate and carbamazepine seemed to be ineffective except in a very small group of truly bipolar children, second generation mood stabilizers have become hugely popular in childhood psychopharmacology. Risperidone, olanzapine, quetiapine and others have been prescribed in significant doses for children diagnosed as childhood bipolar disorder or more recently disruptive mood dysregulation disorder. The drugs clearly reduce rage attacks, irritability and moodiness. However, metabolic side effects including obesity have been extremely troubling. The decision to start a child on second generation antipsychotics should not be taken lightly on the basis of any specific diagnosis. They should be used only briefly in acute psychosocial or family emergencies due to childhood aggressive behaviour. The diagnosis does not give an indication for long-term or lifelong treatment. Short-term use can allow time for parental guidance on behavioural methods of control. The effects of dopamine blockade in the developmental period in animal models on later adult behavior gives much reason for pause [3].

Autism

Autism in its classic sense is a severe disabling neurodevelopmental disorder of language and interpersonal relationships. Usually, it is diagnoseable by age two but in recent years the diagnosis has been broadened to include less severe disorders as autism spectrum disorder (ASD) and these milder cases are sometimes diagnosable only in kindergarten or early school years. Heritability is high and perhaps in about a quarter of the cases full genome scans can find genetic abnormality that may well be a cause of autism in a specific family. However, no one gene mutation is responsible for any significant percentage of severe autism or ASD at the population level. There is no pharmacological treatment of autism.

Many autistic children have repetitive or compulsive behaviours. This component of the illness is responsive to SSRI anti-OCD treatment. The decision to add SSRI in an autistic child should depend on whether the OCD symptoms are a central enough feature of his illness to warrant that medication. It should not be dependent on the physician's need to find something to treat. Autistic children and adolescents do not usually express delusions and hallucinations reminiscent of the positive symptoms of schizophrenia and they do not respond to dopamine D-2 blocking antipsychotic agents. However, if significant aggression or irritability exists low dose antipsychotic medication may give some benefit but overpromising to the parents about psychopharmacological approaches should be avoided [4].

References

1. Belmaker RH, Bleich A. The use of desmopressin in adult enuresis. Mil Med. 1986;151(12):660–2.
2. Robson WL. Clinical practice. Evaluation and management of enuresis. N Engl J Med. 2009;360(14):1429–36.
3. Laporte PP, Matijasevich A, Munhoz TN, Santos IS, Barros AJD, Pine DS, et al. Disruptive mood dysregulation disorder: symptomatic and syndromic thresholds and diagnostic operationalization. J Am Acad Child Adolesc Psychiatry. 2021;60(2):286–95.
4. Lord C, Elsabbagh M, Baird G, Veenstra-Vanderweele J. Autism spectrum disorder. Lancet. 2018;392(10146):508–20.

Chapter 13
Drugs for Alzheimer's and Other Dementia: Are they Worth Anything?

Dementia is a progressive impairment of memory and executive function beginning in the adult, i.e. post developmental period, and usually in old age [1]. It can have many causes including traumatic brain injury, stroke in some parts of the brain, and neurodegenerative disease such as Huntington's or Parkinson's. Vascular disease of the brain deriving from atherosclerosis is also a frequent cause in old age. However, the most common causes are Alzheimer's or frontotemporal dementias. Since these last diseases have a highly increasing incidence with age over 60, the increasing life span in the Western world and now around the globe has led to an exploding prevalence of Alzheimer's and frontotemporal and vascular dementias. These illnesses are often associated with 10–15 years of decreasing ability of self-care and thus impose a huge economic and social burden. Very large amounts of research effort have gone into finding pharmacological or other biological treatments [2].

The usual treatments available for dementia today are the cholinesterase inhibitors.

Acetylcholine is a neurotransmitter used in about 15% of brain synapses. A major but not exclusive source of neurons whose endings release acetylcholine is the nucleus basalis of Meynert. The cholinergic system has long been reported to be involved in memory formation, consolidation and retention in animal studies. Many insecticides work as irreversible acetylcholine esterase inhibitors since insects have prominent cholinergic systems and inhibiting the cholinesterase results in excessive quantities of acetylcholine at the synapse, thus leading to spasm and death of the insect. See Fig. 13.1. When used as insecticides, these compounds are known to be toxic to human beings and mammals as well. When designed for agricultural use, they have to be rapidly degradable in the environment. Human agricultural workers are instructed to use them carefully with gloves and masks. Degradation occurs fast enough so that insect eating birds or other parts of the food chain are usually not affected. It should be noted in these troubled times that irreversible cholinesterase inhibitors with appropriate wind dispersal properties are also used as chemical weapons (nerve gas) by some countries. Defense is based on atropine-like

© The Author(s), under exclusive license to Springer Nature
Switzerland AG 2023
R. H. Belmaker, P. Lichtenberg, *Psychopharmacology Reconsidered*,
https://doi.org/10.1007/978-3-031-40371-2_13

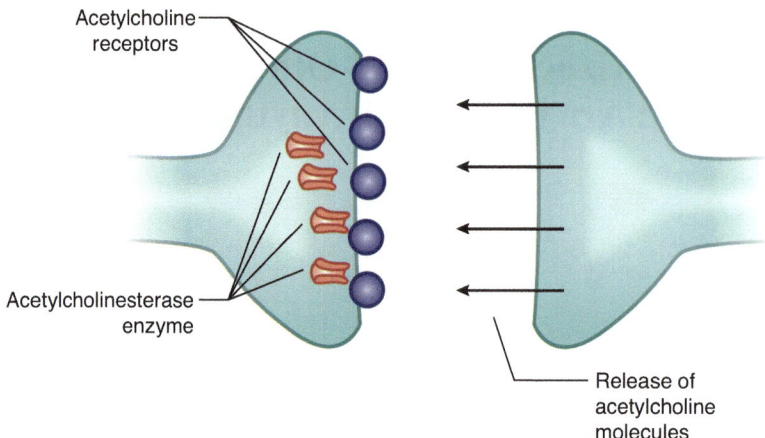

Acetylcholine receptors

Acetylcholinesterase enzyme

Release of acetylcholine molecules

Fig. 13.1   Acetylcholinesterase inhibitors prevent the breakdown of acetylcholine

compounds almost identical to the anti-Parkinsonian anticholinergics used in neurology and psychiatry.

If cholinesterase inhibitors are agricultural poisons, why are we giving them to our elderly loved ones? Tacrine was the first cholinesterase inhibitor to demonstrate memory enhancing properties in Alzheimer patients but this was accompanied by reports of its liver toxicity and other side effects that could be expected from a cholinesterase inhibitor: Since acetylcholine function is increased, the drug causes diarrhea, salivation, constriction of the pupils, sweating and decreased heart rate. Pharmaceutical companies began a race to find safer anti-cholinesterase compounds that might be more specific to acetylcholine esterase in the brain and thus with fewer peripheral side effects. The goal was also to find appropriate half-lives and optimal modes of administration [3].

The most commonly used cholinesterase inhibitor drugs for memory disorder today are donepezil and rivastigmine. These are blockbuster drugs that have sold more than several billion dollars a year in value and are given to hundreds of millions of patients worldwide. Are they effective? Reviews are consistent among all leading neuropsychiatrists that these drugs have a beneficial effect in patients with the memory disorder of Alzheimer's. The effect is of a moderate size but needs importantly to be conceived as equivalent to about 6 months of the average decline experienced by an individual Alzheimer patient. There is no effect on the course of the disease which continues to progress in a deteriorating direction. Thus, if a patient who is diagnosed as suffering from Alzheimer's because over the last year he has ceased to be able to balance his checkbook or remember the law sources that he used to use easily and daily in his work as a lawyer, an anticholinesterase drug might well restore much of both of these functions. However, in 6 months he will have deteriorated to where both of these functions are lost again. If he stops taking the anticholinesterase drug in 6 months, the patient will experience a sudden worsening of symptoms and he will find himself not only unable to manage his checkbook or remember the specific laws that used to be the basis of his profession, but he may forget his own phone number and no longer recognize his grandchild [4].

The clinical value of anticholinesterase drugs in Alzheimer's disease is most salient in the mild to moderate phase of the illness. When used correctly the drugs might allow a patient who would otherwise be unable to enjoy a grandchild's wedding or bar mitzva to be able to do so in a fully oriented manner. However, later in the illness the value of a slight improvement in memory is less likely to translate into an improvement in function and is probably not worth the side effects. These expensive compounds are probably overused because the public does not understand the critical graph shown in Fig. 13.2.

Families are desperate for a cure for this illness and it is difficult for them to accept that a medicine that can help for a while does not help in the long term and does not prevent the gradual decline into a totally dependent and even vegetative state that occurs in most Alzheimer or frontotemporal dementia patients within 5–10 years (sooner in the frontotemporal patients).

Numerous small efforts to increase the efficacy of these compounds have been made, but it also should be clear to physicians that many of the pharmacological adaptations are motivated by the expiration of patents and the desire to promote a new patented variety of the same existing anticholinesterase principle. An example is the use of skin patches for administration of some of these compounds. It is questionable whether the increased cost is worth the benefit and whether these skin patches should be the first line method of administration of anticholinesterase compounds in every patient who needs them.

Another symptomatic drug that improves memory in moderately ill cases of Alzheimer's but does not affect the course of illness is memantine, a drug that probably acts on NMDA receptors rather than cholinergic ones. The target population and clinical effect size are similar to cholinesterase inhibitors [5].

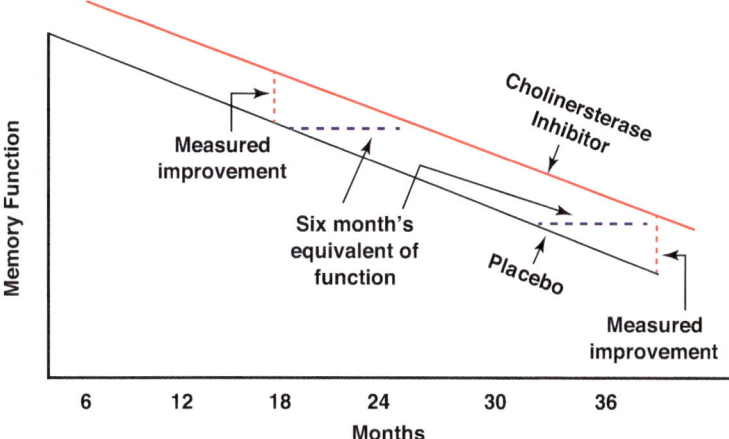

Fig. 13.2 The effect of cholinesterase inhibitors to reverse memory impairment of Alzheimer's with an effect similar to 6 months of disease

Is anticholinesterase treatment an anti-Alzheimer treatment? Absolutely not. Anticholinesterase treatment improves memory in Alzheimer's, in frontotemporal dementia, in vascular dementia, in Parkinson dementia, in Huntington dementia and in dementia from traumatic brain injury as well. Normal controls also show some memory improvement with anticholinesterase inhibition. However, the side effects are such that few students try to use these drugs to improve their exam performance (unlike the stimulant drugs reviewed in Chap. 11 which may or may not improve exam performance but students flock to them because of their euphorigenic effects that give self-confidence and a feeling that the exam was well taken).

The mechanism of the anticholinesterase drugs is clearly via the cholinergic system since anticholinergic drugs impair memory. This latter fact needs to be taken into account when neurologists prescribe anticholinergic drugs for Parkinson's disease or when psychiatrists prescribe anticholinergics for parkinsonian symptoms induced by first generation neuroleptic drug treatment of schizophrenia (see Chap. 6). The confusing use of the term "cholinergic drugs" for both anticholinergic acetylcholine receptor blockers and anticholinesterase inhibitors should be avoided.

Alzheimer's disease may be related to amyloid accumulation in the brain or may have even more complex causes. Considerable research is evaluating whether anti amyloid antibodies may improve Alzheimer's. It is extremely important that psychopharmacology not oversell the small improvement we can achieve symptomatically with anticholinesterase drugs. Patients suffering from this disease and their loved ones should not be misled into thinking that these drugs are a cure. The drugs should be used only for the limited period of time in the illness for which they can give clinically significant, effective relief that is worth their side effects and cost.

Alzheimer and frontotemporal diseases are classified as neurological diseases because their effects in the brain can be demonstrated anatomically, at least in moderately to severely advanced stages of the illness. In early stages of Alzheimer's or frontotemporal dementia CT scans are not always diagnostic. Despite this classification as neurological diseases, a large portion of the care of Alzheimer and other dementia patients falls on the psychiatrist [6]. The patient whose memory is failing him and feels disoriented, unsure of his surroundings and terrified of new experiences will often need psychiatric consultation and psychiatric pharmacologic or non-pharmacologic treatment including environmental planning. It is important that every psychiatrist understand the interaction of the cholinesterase inhibitors with other psychiatric drugs [7] and also the appropriate place of the cholinesterase inhibitors in the proper counseling of the patient and family.

References

1. Arvanitakis Z, Shah RC, Bennett DA. Diagnosis and Management of Dementia: review. JAMA. 2019;322(16):1589–99.
2. Blennow K, de Leon MJ, Zetterberg H. Alzheimer's disease. Lancet. 2006;368(9533):387–403.
3. Joe E, Ringman JM. Cognitive symptoms of Alzheimer's disease: clinical management and prevention. BMJ. 2019;367:l6217.

4. Fan F, Liu H, Shi X, Ai Y, Liu Q, Cheng Y. The efficacy and safety of Alzheimer's disease therapies: an updated umbrella review. J Alzheimers Dis. 2022;85(3):1195–204.
5. Howard R, McShane R, Lindesay J, Ritchie C, Baldwin A, Barber R, et al. Donepezil and memantine for moderate-to-severe Alzheimer's disease. N Engl J Med. 2012;366(10):893–903.
6. Ismail Z, Creese B, Aarsland D, Kales HC, Lyketsos CG, Sweet RA, et al. Psychosis in Alzheimer disease - mechanisms, genetics and therapeutic opportunities. Nat Rev Neurol. 2022;18(3):131–44.
7. Belmaker RH, Bersudsky Y. Lithium-pilocarpine seizures as a model for lithium action in mania. Neurosci Biobehav Rev. 2007;31(6):843–9.

Chapter 14
Drugs for Obsessive Compulsive Disorder: Could Such Obviously Psychological Disorders Have a Pharmacological Treatment?

Obsessive Compulsive Disorder (OCD) involves persistent, recurrent unwanted thoughts and unwanted forced acts, compulsions, that the individual feels compelled to perform and becomes extremely anxious if he tries to resist the urge to perform the compulsive act. OCD was once considered an illness in the family of anxiety disorders, but it is now felt to be a unique condition and this classification fits the unique pharmacology described below. Individual obsessions and compulsions have an eerie psychological quality: For instance, a woman with a young child had a persistent unwanted thought "kill the child" entering her conciousness against her will; she simultaneously felt compelled to hide or throw out all the knives and scissors in the house lest she use them to harm the baby. Often, if not always, the sufferer's personality is such that she would never be at any risk of doing such a thing. Another woman for example, from a conservative religious background, developed an obsession to avoid any motor vehicles and an intrusive thought "if I drive a car I'll have a flat and then some burley truck driver will help me change the flat and I will have sex with him". Another case was a middle aged successful male security worker who had an obsession that entered his mind unwillingly as a feeling "I may have revealed secrets" along with a compulsion to pick up every cigarette butt or other small piece of paper as he walks down the hall if unable or prevented from doing so he developed anxiety, a feeling that he might have written a company secret on the piece of paper that he passed. Needless to say, he took hours to walk even a small distance. A fourth case was an orthodox Jewish patient who began his morning prayers "Blessed are you O Lord" but would suddenly have unwanted thoughts that perhaps he intended the word "Blessed" in its opposite but colloquially used meaning. Therefore, he would need to repeat the blessing compulsively. As you might imagine, fulfilling his daily prayers took him 18 to 20 hours a day and left him with no time to eat or do anything else [1].

OCD rarely leads to hospitalization but its prevalence in the population is higher than schizophrenia or bipolar disorder and it leads to much suffering and disability. The seemingly obvious psychological conflicts in the nature of the obsessions and

R. H. Belmaker, P. Lichtenberg, *Psychopharmacology Reconsidered*, https://doi.org/10.1007/978-3-031-40371-2_14

compulsions led to widespread attempts at psychoanalytic therapy and psychoanalytically oriented psychodynamic therapy. These consistently failed and their failure has been documented by psychoanalysts and psychodynamically oriented therapists themselves. These treatments are no longer indicated. However, behavioral therapy based on psychodynamic understandings was first proposed by Feather and Rhoads [2] in their seminal paper 1974 in the Archives of General Psychiatry and gradually developed into cognitive behavioral therapy and several other approaches that have been proven to have efficacy in most cases of OCD. These therapies are very rarely curative and considerable room exists for combined psychotherapy and pharmacotherapy. Pharmacotherapy alone for OCD is not optimal treatment but often occurs if appropriate behavioral therapists are unavailable or too expensive or if the patient is unwilling to cooperate with a regime of multiple visits for behavior therapy [3].

Yaryura-Tobias [4] first noticed in Europe that some OCD patients being treated for depression responded particularly well in their OCD symptoms as well and particularly with the drug clomipramine (Anafranil), a derivative of imipramine with reuptake inhibiting properties almost entirely limited to the serotonin reuptake system (see Fig. 14.1).

Soon thereafter Zohar and Insel [5] at NIMH in the United States confirmed the finding that clomipramine specifically benefits OCD patients both in their obsessions and in the their compulsions. Zohar and Insel [6] also showed, in an exciting and replicated study, that m-CPP given orally and compared to placebo in OCD patients exacerbated their obsessive-compulsive symptoms. Each patient experienced the exacerbation as a worsening of their own particular obsession or compulsion. There was not an increase in general anxiety. M-CPP administration after a month of treatment with clomipramine revealed that the patients had become relatively resistant to m-CPP exacerbation. M-CPP is a direct agonist at some serotonin

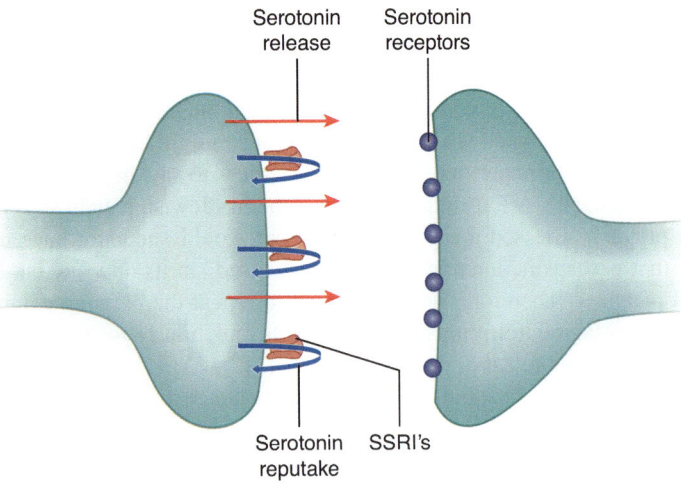

Fig. 14.1 SSRI's work in OCD by inhibiting reuptake pumps for serotonin

(5HT) receptors. Zohar and Insel thus inferred that clomipramine treatment led to a desensitization of serotonin receptors and that the serotonin receptor's activity was somehow abnormally sensitive in the illness of OCD.

Many systems in the body connected with endocrine secretion are mediated by serotonin and can be affected by m-CPP administration. However, none has been found to be abnormal in untreated OCD patients. Post mortem studies of serotonin receptors or CSF studies of 5HT-IAA (5-hydroxyindole acetic acid), the major serotonin metabolite, have not revealed replicable or diagnostic or etiological abnormalities in OCD. MRI imaging of the brain structures or circuits involved in thinking or repetitive action have sometimes been found to be smaller on the average or larger on the average in OCD patients but no imaging examination has been found to reliably and diagnostically distinguish patients from controls or to be useful in the clinical diagnosis of OCD. OCD is a highly heritable illness but no single major gene has been found.

The discovery of clomipramine's effect in OCD by Yaryura-Tobias was serendipitous but the rapid development and study of serotonergic mechanisms in OCD by Zohar and Insel was an example of the phrase "serendipity favors the prepared mind". Moreover, the pharmaceutical companies who had invested so heavily in creating pharmacologically specific serotonin specific reuptake blockers that were hoped to be better treatments for depression, found a completely unexpected surprise reward for their efforts when it became clear that SSRIs were treatments for OCD.

Only serotonin specific reuptake inhibitors worked in OCD among the reuptake inhibitors. Noradrenaline specific reuptake inhibitors are equally efficacious as serotonin specific reuptake inhibitors in the treatment of depression (see Chap. 5 on depression). A true fact, but perhaps less well known, is that serotonin specific reuptake inhibitors are no better but no worse than noradrenaline specific reuptake inhibitors in the treatment of panic disorder (see Chap. 7 on anxiety). However, only in OCD is there a unique specificity for SSRIs. So while non-specificity is a clear theme of "Psychopharmacology Reconsidered" in 2023, the islands of specificity may be beacons that will lead us on to future discovery. The clinician must be acutely careful and aware of those instances when specificity is the proven reality. A patient who responds poorly to a SSRI in depression may be switched to an SNRI or even to bupropion (a dopamine reuptake blocker) as a reasonable clinical next move. For OCD only SSRIs are worth trying or clomipramine, a drug structurally close to imipramine but neurochemically almost an SSRI. Intriguingly, many clinicians feel that clomipramine is the most powerful of the pharmacological treatments of OCD.

Clinical Treatment of OCD

While depression typically has a statistically and clinically significant response to antidepressant treatment by 3 weeks, response of OCD to SSRIs most often takes about 6 weeks and can continue to have increasing benefit even after 6 weeks. Doses used for the treatment of OCD are higher than those used for depression and doses often must be raised to levels that cause significant side effects to achieve clinical benefit. Most patients, but not all, suffer so much from OCD that they are willing to tolerate the side effects [7].

Side Effects of SSRI Treatment of OCD

The doses of clomipramine or SSRIs necessary to effectively treat OCD almost invariably induce sexual dysfunction, particularly delayed orgasm in the male or female. Many treatments have been suggested for this side effect which is troublesome for many patients but not all. It is important the physician have the training to maintain an open and effective discussion on this matter without being judgmental toward those patients for whom it is an important side effect and to be equally nonjudgmental to those patients who could not care less. Numerous pharmacological remedies have been proposed in case reports for this side effect including small doses of serotonin receptor blockers taken on an as needed basis an hour or two before sexual activity. Most of these recommendations have not stood up in the test of controlled studies but the controlled studies might be sampling a different patient population and the clinician who works with OCD patients should familiarize himself with all of the possible treatments for the side effects that could otherwise impair adherence to treatment. Sildenafil or other PDE-5 inhibitors, while not approved for this indication, are often reported anecdotally as increasing sexual responsivity to orgasm in both males and perhaps in females and may be taken 1 h before intercourse in patients having difficulty reaching orgasm because of ant-OCD treatment.

Surprisingly, small doses of dopamine receptor blockers including the classics such as chlorpromazine, but more often nowadays aripiprazole or risperidone, augments the anti-OCD effects of SSRIs in severe cases. Only small doses should be used as OCD patients can become oppositional to the antipsychotic drugs' side effects and eventually lose adherence to their whole treatment program. Co-morbid tic disorder or Tourette's syndrome of course provide a conceptual handle for neuroleptic augmentation. Lithium, which augments antidepressant treatment in some antidepressant nonresponders in depression (see Chap. 11 on mood stabilizers), does not seem to reliably augment the anti-OCD effect of SSRIs [8].

OCD Symptoms in Other Syndromes and in Childhood

OCD symptoms may be prominent in some autistic children and evidence suggests that these symptoms respond to SSRI's. However, stereotypies alone probably do not justify clinical enthusiasm for this window into the overall treatment of autism. Some patients with schizophrenia have prominent OCD symptoms and have been called "schizo-OCD". OCD symptoms in such patients with schizophrenia respond to SSRI, although the risk of exacerbation of psychosis has not been ruled out in the scientific literature or clinical practice [9].

OCD symptoms can occur in a benign manner in childhood from age 7 to age 12 with no prognostic implications; however, the full illness of OCD can also begin in childhood and responds to treatments as in the adult form. OCD in all its classic symptoms can occur as a sequella of childhood streptococcal infection, just as childhood chorea or tic disorder sometimes do. These have been called PANDAS (Pediatric Autoimmune Neuropsychiatric Disorders Associated with Streptococcal Infections) and have strengthened speculation on the role of the basal ganglia in OCD in general . Use of penicillin in treatment is experimental. OCD along with depression and anxiety seems to have increased in the isolation of the corona period and specific treatments for these cases is not yet determined [10] .

Obsessive-compulsive personality disorder is a clinically very real, prevalent and disruptive facet of human behavior with implications for marital, occupational, and societal adjustment. It does not respond to treatment with SSRI [11].

OCD is as Churchill said in a different context "an enigma wrapped in a mystery". The patient's complaints seem so psychological that the clinician is easily deluded into thinking that psychodynamic interpretations will be of help. There is probably no psychiatrist or psychologist who has not fallen into this trap at least a few times. Similarly, it is likely that no psychiatrist or psychologist has ever succeeded in helping an OCD patient with one of these obvious interpretations. It is also of no help to the patient and society to quote head-to-head controlled studies as to whether behavior therapy is better, equal to or less good than pharmacotherapy in OCD. The two treatments should optimally be combined but medicine is the realm of the possible and combination can only be achieved in those patients who have access to and can tolerate both therapies in real life [12]. OCD often waxes and wanes throughout a lifetime course, and it does not necessarily require continuous or prophylactic treatment either pharmacological or behavioral during periods when it is in remission.

Clinical Vignettes

1. A 20 year old young woman had been an outstanding high school student but after completion developed severe obsessional thoughts about forbidden foods popping out of her mouth into other people's mouths, leading her to avoid any food preparation and to restrict her own foods to a very limited variety that might not pop out of her mouth. Her weight was normal and after a few months she had also become depressed. Treatment with 20 mg of fluoxetine relieved her depression and partially relieved her OCD after 6 weeks. Increase of dosage to 40 mg relieved over 80% of her OCD symptoms. She refused numerous attempts to engage her in behavioral psychotherapy.

2. A 67 year old man has been the economist at a major government bureau with a lifelong history of bipolar disorder well controlled with lithium and never requiring hospitalization. He had an obsessive compulsive personality disorder his whole life. In his early 60 s he developed OCD with repeated thoughts that he would be unable to urinate thus leading to repeated and almost continual going to the bathroom to attempt to urinate. Urological exam was normal. Addition of 20 mg fluoxetine to his mood stabilizer treatment led within a few weeks to a full blown manic attack.

References

1. Grant JE. Clinical practice: obsessive-compulsive disorder. N Engl J Med. 2014;371(7):646–53.
2. Rhoads JM, Feather BW. The application of psychodynamics to behavior therapy. Am J Psychiatry. 1974;131(1):17–20.
3. Fux M, Levine J, Aviv A, Belmaker RH. Inositol treatment of obsessive-compulsive disorder. Am J Psychiatry. 1996;153(9):1219–21.
4. Volavka J, Neziroglu F, Yaryura-Tobias JA. Clomipramine and imipramine in obsessive-compulsive disorder. Psychiatry Res. 1985;14(1):85–93.
5. Zohar J, Insel TR. Drug treatment of obsessive-compulsive disorder. J Affect Disord. 1987;13(2):193–202.
6. Zohar J, Insel TR, Zohar-Kadouch RC, Hill JL, Murphy DL. Serotonergic responsivity in obsessive-compulsive disorder. Effects of chronic clomipramine treatment. Arch Gen Psychiatry. 1988;45(2):167–72.
7. Skapinakis P, Caldwell DM, Hollingworth W, Bryden P, Fineberg NA, Salkovskis P, et al. Pharmacological and psychotherapeutic interventions for management of obsessive-compulsive disorder in adults: a systematic review and network meta-analysis. Lancet Psychiatry. 2016;3(8):730–9.
8. Abramowitz JS, Taylor S, McKay D. Obsessive-compulsive disorder. Lancet. 2009;374(9688):491–9.
9. Goodman WK, Storch EA, Sheth SA. Harmonizing the neurobiology and treatment of obsessive-compulsive disorder. Am J Psychiatry. 2021;178(1):17–29.
10. Pan KY, Kok AAL, Eikelenboom M, Horsfall M, Jörg F, Luteijn RA, et al. The mental health impact of the COVID-19 pandemic on people with and without depressive, anxiety, or obsessive-compulsive disorders: a longitudinal study of three Dutch case-control cohorts. Lancet Psychiatry. 2021;8(2):121–9.
11. Hirschtritt ME, Bloch MH, Mathews CA. Obsessive-compulsive disorder: advances in diagnosis and treatment. JAMA. 2017;317(13):1358–67.
12. Stein DJ, Costa DLC, Lochner C, Miguel EC, Reddy YCJ, Shavitt RG, et al. Obsessive-compulsive disorder. Nat Rev Dis Primers. 2019;5(1):52.

Chapter 15
Cannabis: A Useful Psychotropic for Pain, PTSD and Sleep or a Gateway to Schizophrenia?

Cannabis refers to substances derived from the hemp plant plant cannabis sativa and its various strains or species. The plant has been used since ancient times throughout much of the world both for the production of rope and for the psychoactive effects achieved usually by smoking the leaves or flower buds [1] . The effects of cannabis in humans include relaxation, sense of well-being, enhancement of sexual pleasure, enhancement of pleasure from food and increased appetite and in the appropriate setting increased ease in falling asleep. In social settings cannabis ingestion enhances talkativeness and desire for social interaction. It impairs reactivity to distracting stimuli and impairs driving ability [2]. In higher doses a change in conscious state may be induced with elation or spontaneous giggling. In even higher doses it induces feelings of suspiciousness and paranoia and in even higher doses causes an acute toxic psychosis that may resemble delirium. Since the drug is usually delivered by smoking, it is hard to control an exact dose. Oral delivery of this oily compound in edibles is even harder to dose, since food or other factors may cause delayed absorption and consumption of additional doses before an original dose is absorbed, thus leading to overdose.

The extraction of cannabis to enhance the concentration of active ingredients has gone on for many years but has accelerated in recent decades. Strains of cannabis sativa, the natural weed, have been selected for increasing concentrations of psychoactive substances leading to effects of the drug ingested clinically or recreationally that are increasingly different from those experienced by humanity in the past.

The active ingredient of cannabis is tetrahydrocannabinol (THC) but almost 150 other cannabinoid (CB) compounds with possible psychoactive effects exist in the plant and its extracts. The second most common cannabinoid in the plant after THC is cannabidiol (CBD) which will also be discussed below. THC was first identified as the active component of cannabis by Raphael Mechoulam, an Israeli scientist. Its receptor in the brain, the cannabinoid receptor, was identified soon afterwards. The endogenous ligand of this system was then discovered and called anandamide, from a Sanskrit word meaning "bliss". This chronology from an exogenous compound or

R. H. Belmaker, P. Lichtenberg, *Psychopharmacology Reconsidered*, https://doi.org/10.1007/978-3-031-40371-2_15

drug to the discovery of its receptor to the final discovery of a new neurochemical system is a common theme in the history of psychopharmacology. Unlike the dopamine, serotonin, noradrenaline, GABA or acetylcholine systems, the anandamide system exists diffusely throughout the brain and the peripheral body and is difficult to locate to specific functions or pathways. There are two receptors: CB1 receptors are found mostly in the brain and CB2 receptors mostly in the peripheral immune system. See Fig. 15.1 for a simplified version of anandamide's role, using the GABA receptor only as an example. As in Fig. 15.1, anandamide is often secreted by postsynaptic cells to receptors on presynaptic cells and thereby modify the effect of other neurotransmitters. It is chemically neither a monoamine, nor an amino acid nor a peptide but a fatty acid related to arachidonic acid. Anandamide can function not as a simple neurotransmitter conveying messages from one neuron to another but as a modulatory neurohormone. Many of its neurochemical effects are still difficult to define and it may have numerous effects that are as yet undiscovered [3].

THC is widely used as a recreational substance by persons seeking its relaxing and pleasure enhancing qualities. Most studies find that as such a substance it causes considerably less physical and social damage than alcohol or nicotine, two other widely used self-administered legal psychoactive compounds [4]. However, THC use is a moving target. It is difficult to apply studies that were done as recently as 5 years ago using the concentrations available for smoking cannabis then with the much more concentrated forms of THC that can be smoked today from highly selected plant strains. Thus, many in the field support a regulated approach where pharmacies or government supervised delivery stores would supply and sell products whose THC concentration is known and can be controlled [5].

Medical indications for THC are few and stand in direct contrast to the extensive claims made for huge therapeutic properties for this compound. THC is useful in

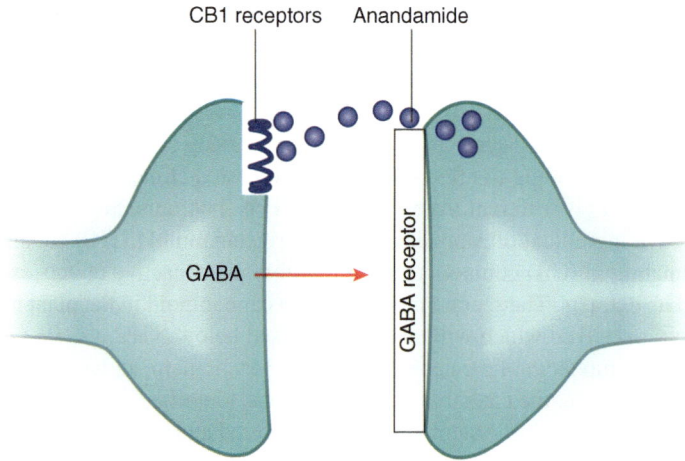

Fig. 15.1 Anandamide can allow feedback from the postsynaptic GABA cell via presynaptic CB1 receptors

chemotherapy induced emesis and probably useful in the severe insomnia of PTSD patients induced by their nightmares. It is useful as an appetite stimulant in patients with AIDS. It is somewhat effective in neuropathic pain such as in diabetes. It is poor as an analgesic for acute orthopaedic pain, in trauma pain, in myocardial infarction or acute pain of other sources, especially if compared to opiate analgesics [6]. Despite its tendency to increase the pleasure of eating and induce weight gain in recreational users, it has not been found to be useful in anorexia nervosa. Its use for depression or anxiety has not been supported by controlled trials [7]. It also does not seem to have overall benefits in the PTSD syndrome other than to alleviate the nightmare-induced insomnia. THC is well proven to worsen psychosis and exacerbate episodes in bipolar disorder [8]. Thus, it is far from a panacea and its exploding use recreationally around the world will provide much employment for psychiatrists treating its various psychiatric complications, just as alcohol use has caused mild enjoyment to many but liver cirrhosis, heart disease and esophageal cancer to many others.

THC derives from a common weed and in this sense is a "natural substance". Its evolutionary purpose in the plant was probably to avoid insect predation by interfering with insect behavior. However, opium which is derived from the poppy plant is also natural as is digitalis which is derived from foxglove and atropine with is derived from belladonna. All of these natural substances can be deadly and misused. The statement that marijuana is natural and is therefore somehow harmless is highly misleading and alarms any thoughtful psychopharmacologist [9]. It is painful to hear a patient with psychosis reject a dopamine receptor blocker treatment as "unnatural" while using daily doses of marijuana which exacerbates his psychosis but which he steadfastly believes is "natural". Perhaps even more worrisome is the common observation that casual use of marijuana may induce a missionary-like belief in the substance. The belief system is certainly part of the human brain and mind and THC may somehow tap into this system and convince users that this compound is somehow far more beneficial than it objectively is [10].

The low potency exposure to THC that existed with mankind in recent millennia rarely if ever led to a clear addiction syndrome. This is absolutely not true today. Increasingly patients are presenting with daily use of high potency THC based compounds and are unable to stop their use even though it has come to totally interfere with their function in work or family. The THC addiction syndrome includes multiple daily use, inability to stop even when desiring to do so and serious disruption of commitments to family and work [11]. An amotivational syndrome sometimes develops and cognitive impairment may occur. Acute withdrawal symptoms include anxiety, insomnia and generalized discomfort but carry no danger like the delirium tremens of alcohol withdrawal. No hospitalization is necessary for withdrawal but specialized facilities can help support recovery from marijuana addiction. Chronic psychosis after chronic marijuana use occurs rarely.

The highest danger for behavioural and brain damage due to marijuana use appears to be in the human developmental period. Pregnant women must be strongly advised against use of marijuana despite their occasional belief that this is a "natural" substance [12]. Teenagers with developing brains are also at high vulnerability and frequent use of marijuana in the teenage years has been incontrovertibly associated with elevated rates of psychosis later on in life [13].

The second most common compound in the marijuana plant is cannabidiol (CBD). CBD is not euphorigenic and all current evidence suggests that it is not helpful in anxiety, depression or in sleep. It is apparently effective therapeutically in some rare forms of childhood epilepsy [14]. While there were some enthusiastic studies in autism, no reliable positive data has emerged. There is some epidemiological data to suggest that the combination of THC with CBD in many strains of cannabis sativa makes that strain less likely to induce psychosis. From this observation some researchers have been studying CBD as an antipsychotic. Results are preliminary and do not justify routine clinical use in place of dopamine receptor blockers when indicated.

CBD is widely administered by physicians around the world most probably with placebo effects only. It is also available over the counter in many countries in oily drops. Sometimes CBD mixed with THC is available in variable ratios and is given for illnesses such as fibromyalgia and borderline personality disorder for which no other pharmacological treatments exist. The ethics and dangers of such use of CBD should give the psychiatric profession and the general public considerable cause for thought [15].

Rimonabant and Dronabinol

Rimonabant is an inverse agonist at CB1 receptors and has appetite suppressant properties. It was marketed in several countries for obesity treatment but rejected by the FDA because of depressogenic and suicidogenic risks. Since THC enhances appetite and is an agonist at the CB1 receptor, the development of rimonabant was rational. The psychiatric side effects underscore the potency of intervening in this system, both with THC or with its blockers. Rimonabant is not available for nor proven to be effective as a treatment for marijuana intoxication or addiction (for contrast, see methadone in opiate chapter). Dronabinol is THC in sesame oil in capsules approved for chemotherapy-induced emesis and HIV induced anorexia in several countries. There have been few reports of dronabinol abuse. Some claim that marijuana whole plant derivatives are superior because there is a "symphony" of interactive effects of the numerous cannabinoids in the plant. This claim is unproven and reminiscent of many unproven claims of marijuana enthusiasts and activists.

Synthetic Cannabinoids

A plethora of street drugs are appearing yearly which are potent agonists at CB1 and perhaps other receptors as well. Some of these "escaped" from research laboratories where they were being studied for specific effects on CB1 receptors; others are loosely related chemically to THC and were synthesized for the purpose of street use. Many are called "synthetic marijuana" and are sprayed on leaves or buds for illegal use. Many are potent psychotogenic agents even if they have some similarities to THC; they cause considerable psychiatric morbidity and are a threat to public health, no matter what legal policy emerges regarding marijuana itself.

References

1. Volkow ND, Swanson JM, Evins AE, DeLisi LE, Meier MH, Gonzalez R, et al. Effects of cannabis use on human behavior, including cognition, motivation, and psychosis: a review. JAMA Psychiatry. 2016;73(3):292–7.
2. Brubacher JR, Chan H, Erdelyi S, Staples JA, Asbridge M, Mann RE. Cannabis legalization and detection of tetrahydrocannabinol in injured drivers. N Engl J Med. 2022;386(2):148–56.
3. Albaugh MD, Ottino-Gonzalez J, Sidwell A, Lepage C, Juliano A, Owens MM, et al. Association of Cannabis use during Adolescence with Neurodevelopment. JAMA Psychiatry. 2021;78(9):1–11.
4. Englund A, Freeman TP, Murray RM, McGuire P. Can we make cannabis safer? Lancet Psychiatry. 2017;4(8):643–8.
5. Hall W, Degenhardt L. Adverse health effects of non-medical cannabis use. Lancet. 2009;374(9698):1383–91.
6. Campbell G, Hall WD, Peacock A, Lintzeris N, Bruno R, Larance B, et al. Effect of cannabis use in people with chronic non-cancer pain prescribed opioids: findings from a 4-year prospective cohort study. Lancet Public Health. 2018;3(7):e341–e50.
7. Black N, Stockings E, Campbell G, Tran LT, Zagic D, Hall WD, et al. Cannabinoids for the treatment of mental disorders and symptoms of mental disorders: a systematic review and meta-analysis. Lancet Psychiatry. 2019;6(12):995–1010.
8. Di Forti M, Quattrone D, Freeman TP, Tripoli G, Gayer-Anderson C, Quigley H, et al. The contribution of cannabis use to variation in the incidence of psychotic disorder across Europe (EU-GEI): a multicentre case-control study. Lancet Psychiatry. 2019;6(5):427–36.
9. Gobbi G, Atkin T, Zytynski T, Wang S, Askari S, Boruff J, et al. Association of Cannabis use in adolescence and risk of depression, anxiety, and suicidality in young adulthood: a systematic review and meta-analysis. JAMA Psychiatry. 2019;76(4):426–34.
10. Richter KP, Levy S. Big marijuana--lessons from big tobacco. N Engl J Med. 2014;371(5):399–401.
11. Hindley G, Beck K, Borgan F, Ginestet CE, McCutcheon R, Kleinloog D, et al. Psychiatric symptoms caused by cannabis constituents: a systematic review and meta-analysis. Lancet Psychiatry. 2020;7(4):344–53.
12. Paul SE, Hatoum AS, Fine JD, Johnson EC, Hansen I, Karcher NR, et al. Associations between prenatal cannabis exposure and childhood outcomes: results from the ABCD study. JAMA Psychiatry. 2021;78(1):64–76.
13. Wilson J, Freeman TP, Mackie CJ. Effects of increasing cannabis potency on adolescent health. Lancet Child Adolesc Health. 2019;3(2):121–8.
14. Friedman D, Devinsky O. Cannabinoids in the treatment of epilepsy. N Engl J Med. 2015;373(11):1048–58.
15. Moore TH, Zammit S, Lingford-Hughes A, Barnes TR, Jones PB, Burke M, et al. Cannabis use and risk of psychotic or affective mental health outcomes: a systematic review. Lancet. 2007;370(9584):319–28.

Chapter 16
Ketamine and Psychedelics in PTSD and Depression: The New Frontier or the New Pandemic?

Introduction

The media and internet are flooded with news about psychedelic compounds previously illegal but widely used recreationally that are now being investigated for therapeutic potential. We will review the status of four major compounds in this group, distinguishing possible therapeutic use in depression and PTSD from unproven risk-benefit ratio for unregulated recreational use.

Ketamine

Ketamine was developed as a general anaesthetic with minimal respiratory depression properties for use in short anaesthetic procedures. However, patients complained that they had memory of events that occurred during anaesthesia and also unpleasant hallucinations that were remembered as well. It has now less general anaesthetic use except for veterinary use and paediatric use where the complaints have been fewer. Ketamine also appeared in the street market as a drug of abuse where it is reported to give hallucinations and psychedelic experiences [1] .

The chemical parent of ketamine, phencyclidine, was widely used illegally as a psychotomimetic and in animal models it became the basis of the glutamate (NMDA) theory of psychosis. This theory has been heuristic. However, as sceptical psychopharmacologists, we must approach the antidepressant actions of ketamine and glutamate theories of its antidepressant action while knowing its parallel life as a psychotomimetic (See Chap. 11 on stimulants for an example of parallel contradictory theories in psychopharmacology).

Early serendipitous reports that ketamine elevated mood led to a small clinical controlled trial finding that subanesthetic doses of intravenous ketamine had a

R. H. Belmaker, P. Lichtenberg, *Psychopharmacology Reconsidered*, https://doi.org/10.1007/978-3-031-40371-2_16

dramatic and rapid effect to improve mood. Many of the volunteers may have had previous experience with illegally obtained street ketamine; however the rapid effect on depression and suicidality seemed far beyond what would be expected of the usual street drug. The value of the unexpected finding was connected to recent work on the NMDA receptor, one of the receptors of glutamate in the brain. Glutamate is an excitatory neurotransmitter with a very complex receptor (see Fig. 16.1). Since the monoamine synapse had been yielding less and less antidepressant fruit in recent years, the glutamate system was drawing increasing interest as a possible frontier for antidepressant work. Behavioural models in animals suggested that glutamatergic agonists might be antidepressant and the ketamine findings as a clinical antidepressant fit the story.

Since that initial report numerous studies have been carried out. The effect of ketamine was first studied as an intravenous single infusion in a laboratory study with carefully controlled dosing to minimize the number of patients who would have psychedelic symptoms such as out of body experiences, hallucinations or derealization. The therapeutic index was narrow and care was necessary but early investigators felt that the antidepressant effect could clearly be relegated to lower doses than the psychedelic effect of ketamine. The antidepressant effect and particularly the anti-suicidal effect seemed particularly well suited to the emergency room situation where large number of patients arrive every year in acute suicidal states [2]. These states are frightening and dangerous to both patients, families and staff and have often resulted in closed ward hospitalizations for long periods of time. Early reports suggested that ketamine could resolve acute suicidality within hours. Studies ensued to determine whether the effect could be long lived. Present data suggests that a certain percentage of patients have a long lasting effect whereas others require repeat administration.

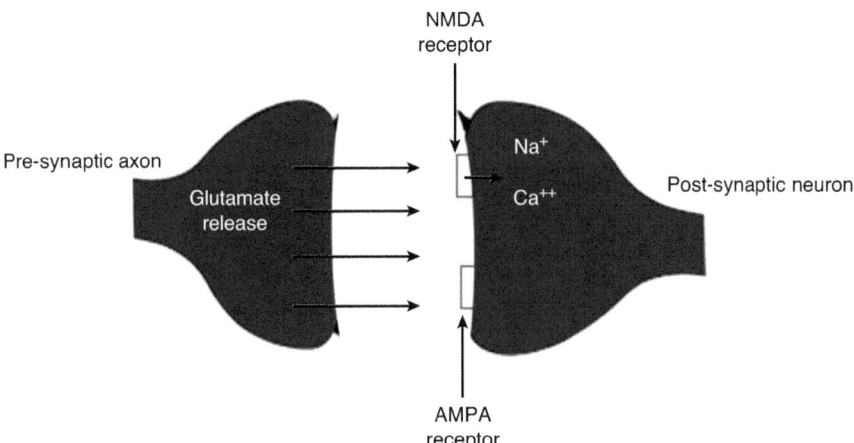

Fig. 16.1 The NMDA receptor is one part of the receptor system for glutamate neurotransmission

In order to facilitate the rapid development of the ketamine concept for clinical use, the FDA granted Johnson & Johnson orphan drug status for its me-too single isomer derivative of ketamine. Ketamine is a racemic mixture of s- and l- ketamine and from it Johnson & Johnson extracted the s-ketamine and began clinical trials with s-ketamine called esketamine. Esketamine was brought to market for the treatment of depression in 2020 and nasal esketamine for chronic resistant depression has been approved in several countries since then. The use has been restricted so far to patients resistant to standard antidepressant therapy and some abuse has been reported. The effect is not always so rapid that it occurs after a single dose but it does not seem to be dependent on a concurrent psychotherapeutic setting as in some of the drugs discussed below but this has not yet been fully explored [3].

MDMA (Ecstasy)

MDMA is a party drug that induces an out of sense experience including general love of mankind according to reports of many users. Early reports that it destroyed serotonergic cells irreversibly were not replicated in human relevant dosing. Use of MDMA spread rapidly in some circles around the world and some deaths were reported because of its effect to impair temperature regulation if it were used by dancers during all night parties in hot climates or closed rooms. Clinical trial and use of MDMA was promoted by the MAPS foundation, a private group headed by Rick Doblin who conducted clinical trials beginning with studies of PTSD. These studies have generally given quite positive results. The MDMA is administered one or two sessions separated by a week or two and accompanied by almost 24 h of psychological encouragement, debriefing and support for the patient to achieve an out of self experience in which he can see his symptoms or trauma as curable and small in relation to the immenseness of the universe. The coexistence of MDMA in therapeutic centers with its continued street use makes for some problems. It is our clinical experience that some patients who benefited from participation in clinical trials went on to repeated street use of the drug in an uncontrolled fashion that was damaging to them and their families [4].

Psilocybin

Psilocybin has been used for millennia especially in the Americas by native peoples who derived it from mushrooms and used it in carefully controlled religious ceremonies where it induced spiritual feelings, out of body sensations and peacefulness. For modern psychopharmacology wholly synthesized psilocybin or its derivatives are used. A most recent controlled trial in the NEJM 2022 found a clear dose repose relationship where 1 mg and 10 mg were ineffective but 25 mg had a marked effect to relieve depressive symptoms . The administration of the psilocybin was similar to that of MDMA reported above in that the psychedelic experience was created for one or two sessions along with massive counselling, support and abreaction with no intention to use the drug on a daily or chronic basis [5].

LSD

LDS (Lysergic acid diethylamide) was first synthesized by Hoffman in the 1930s and its use popularized by Timothy Leary in the 1950s. Its ingestion leads to hallucinations both in the auditory and visual sphere that can last for hours or up to a day. The hallucinations and perceptual distortions can be amazingly complex and include cultural background such as hearing a Beethoven's symphony or fantasy such as seeing crystal structures that do not exist in reality or in the patient's known history. The perceptual distortions do not in themselves have a euphoric or antidepressant component but more seem to take place on a gradient from frightening to interesting. Novice users could find the drug frighting if not prepared but experienced users seem to use the drug to relieve boredom or expand their horizons in some sense. A powerful phenomenon often occurs, as described above for cannabis, where users become missionaries for the drug and recommend it to other not only because of financial incentives to resell or commercialize the drug but because of the nature of how it affects their experiences. The drug cannot be used to overdose and it is not addicting in the manner of opiates. Some investigators have felt since the 1950's that the LSD experience can be used as an adjunct to intensive psychotherapy to help patients overcome PTSD and depression by enhancing their awareness of positive sensation, connectedness to other humans, or the humility of our place in the universe, all cognitive phenomena that have biological correlates perhaps affected by the drug [6].

Mechanisms

The substances described above are all different from classic psychotropic compounds because they are not used on a daily basis with a clear chronologic relationship between symptom amelioration and length of exposure to the drug [7]. Moreover, usually the drug is administered in a special intensely psychotherapeutic setting and the drug is viewed as a method of entry into a special receptive psychotherapeutic state rather than as a psychopharmaceutical in the classic sense. While ketamine exerts its major effects on NMDA receptors for the neurotransmitter glutamate, psilocybin seems to affect several specific subtypes of serotonin receptors and LSD is clearly known to exert its hallucinogenic and psychedelic properties via stimulation of one subtype of the serotonin 5HT-2 receptor. MDMA (3,4-methylen edioxy-methamphetamine) has several psychopharmacological actions but is closest to cocaine and amphetamine in releasing several monoamines simultaneously in the synapse.

These new compounds have been hailed in the popular press as representing the beginning of a new era in psychopharmacology. Those who used the drugs recreationally since the 1950's feel vindicated, but there is no proof in the medical research literature so far that these drugs are safe for general use and have a risk-benefit ratio appropriate for recreational use. Caution would be the better part of valor for our situation with these compounds [8].

References

1. Sanacora G, Frye MA, McDonald W, Mathew SJ, Turner MS, Schatzberg AF, et al. A consensus statement on the use of ketamine in the treatment of mood disorders. JAMA Psychiatry. 2017;74(4):399–405.
2. Abbar M, Demattei C, El-Hage W, Llorca PM, Samalin L, Demaricourt P, et al. Ketamine for the acute treatment of severe suicidal ideation: double blind, randomised placebo controlled trial. BMJ. 2022;376:e067194.
3. Smith-Apeldoorn SY, Veraart JK, Spijker J, Kamphuis J, Schoevers RA. Maintenance ketamine treatment for depression: a systematic review of efficacy, safety, and tolerability. Lancet Psychiatry. 2022;9(11):907–21.
4. Mithoefer MC, Grob CS, Brewerton TD. Novel psychopharmacological therapies for psychiatric disorders: psilocybin and MDMA. Lancet Psychiatry. 2016;3(5):481–8.
5. Madras BK. Psilocybin in treatment-resistant depression. N Engl J Med. 2022;387(18):1708–9.
6. Davis AK, Barrett FS, So S, Gukasyan N, Swift TC, Griffiths RR. Development of the psychological insight questionnaire among a sample of people who have consumed psilocybin or LSD. J Psychopharmacol. 2021;35(4):437–46.
7. Doss MK, Madden MB, Gaddis A, Nebel MB, Griffiths RR, Mathur BN, et al. Models of psychedelic drug action: modulation of cortical-subcortical circuits. Brain. 2022;145(2):441–56.
8. Johnson MW, Hendricks PS, Barrett FS, Griffiths RR. Classic psychedelics: an integrative review of epidemiology, therapeutics, mystical experience, and brain network function. Pharmacol Ther. 2019;197:83–102.

Chapter 17
The Placebo Response in Psychopharmacology and the Use of Nutraceuticals in Clinical Psychopharmacology

The Placebo Response in Psychopharmacology

We have been careful in this textbook to limit our purview to matters of psychopharmacology. Writing about the placebo effect, its meaning and relevance to our work, challenges that stricture, as we will try to make clear in this chapter. The placebo effect has become far more than a methodological nuisance for researchers trying to assess a drug's efficacy. In the words of the social psychologist W.J McGuire, there are three stages in the life of an artifact: "First it is ignored, then it is controlled for its presumed contaminating effects, and finally it is studied as an important phenomenon in its own right." We are by now deep into the third stage. And a psychiatrist seeking literacy in the nuances of psychopharmacology, its benefits, harms and limitations, should possess a subtle understanding of the complex area of "placebo".

A patient – probably in the context of a research protocol – receives a sugar pill for her depression, and her condition improves. The possible reasons for this are many, including:

1. the natural course of the illness,
2. various aspects of standard care which accompany the particular therapeutic intervention, such as nursing care, healthy dieting, etc.,
3. easing of anxiety by diagnosis and treatment;
4. the support of the concerned family,
5. regression to the mean, a statistical phenomenon whereby if one receives treatment at a particularly bad time of suffering, then randomly his suffering can be expected to moderate and improve,
6. The Hawthorne effect, which is the tendency of people to improve or perform better when they know that they are being watched [1],
7. The conscious expectation of improvement aroused by treatment, in all its aspects: a prescribing physician who appears to understand the problem and offers a solution,

© The Author(s), under exclusive license to Springer Nature Switzerland AG 2023
R. H. Belmaker, P. Lichtenberg, *Psychopharmacology Reconsidered*,
https://doi.org/10.1007/978-3-031-40371-2_17

8. The conditioned response to the act of receiving treatment; this does not involve conscious awareness,

It is important to realize that only the last two factors should be considered part of the placebo effect. That means that many placebo responders are actually getting better unrelated to the fact of their having ingested a placebo. Though this led some to wonder whether the placebo effect is not merely a myth [2], the strong consensus is that the effect is real, as confirmed by different findings:

1. A wide variety of placebos are all more effective than receiving no pill, as on a waiting list control group [3].
2. Recent years have found a burgeoning of research on open label placebos, where research participants are told that they are receiving a placebo, and where the outcome is then compared with groups receiving no pill but equal attention. The placebo group proves superior [4, 5].
3. Complex and elegant experiments have shown that improvement with placebo can be pharmacologically blocked, in a manner determined by how the effect of the placebo – i.e. the expectation or conditioning for improvement – was achieved [6]. More recent neuroimaging studies have shown how placebo-produced positive expectation can have specific neurobiological effects, such as activating the endogenous opioid system [7].

The placebo might be considered the oldest, most ubiquitous treatment in the history and prehistory of medicine. Tribal rituals, shamans and healers, blood-letting and a panoply of herbs and other substances, could provide relief only in the context of society offering hope and trusting the authority of traditions and those offering these remedies. Indeed, one might reasonably argue that the fruit eaten in the Garden of Eden story, whose potency was touted by the Deity and amplified by the snake, was amongst many other meanings a primal tale of placebo effect.

There is no clear definition of the placebo effect, and the variety of methods used to arouse it in a research context – and probably even moreso in clinical interactions – is as varied as the imagination of the practitioners. A very partial list includes dummy pills, relaxation, idle chatter, sham acupuncture, vitamins, and subtherapeutic dosages of medicines.

The common denominator of all these placebos is that they act in a top-down direction in order to produce their effects. That is to say, a person interacts with another person, in such a manner that she produces an expectation in a culturally meaningful way, or perhaps activates a conditioned response, which leads to relief from suffering. The evidence for this relief is not limited to subjective self-reporting, but as noted can also be physiologically verified and, under proper conditions, pharmacologically blocked. And while this effect has been most studied in conditions where subjective rating is the key – depression, anxiety, pain – placebo effects have also been demonstrated conclusively for physical ailments, such as Parkinson's disease.

The placebo, then, can be defined by its top-down mode of action. One may protest that if so, we relegate psychotherapy to the realm of placebo therapy. To this we respond in two ways. First of all, once we understand the top-down nature of place-bos, its potential role in ameliorating suffering, and the complex manipulations by which one might cause such an effect to happen, then it is not to bury psychotherapy but to praise the placebo that we consider the former an example of the latter. This is particularly so for those who interpret the placebo effect as a "meaning effect", which is to say that the cultural value of the various rituals and actions involved are the source of positive change [8]. Secondly, referring to placebos as a top-down interventions means that, when you think about it, all psychosocial modalities can be considered examples of exploiting the placebo effect. Justifiably, some have argued for abolishing the very term "placebo", with its hint of quackery and implied denigration of a universal human phenomenon. The placebo effect is a challenge to a narrowly mechanistic approach to treatment, and demands that we consider fac-tors in our treatment beyond neurochemical effects upon the brain.

By contrast, pharmacological treatments are invariably bottom-up interventions: by manipulating a receptor, or second messenger, or other activity at the cellular or tissue level, we aim to produce an improvement in the wellbeing of the person. The clinician must understand that, though operating in opposite directions, these approaches are not mutually exclusive; indeed, working together, they can synergis-tically provide the greatest benefit. At least one study involving many treating psy-chiatrists prescribing different medications has suggested that more of the variability in outcome derives from the psychiatrist than from the pill [9]. All this is but another way to argue for the necessity of an intricate biopsychosocial approach in treatment, which is no less relevant when we use medication.

Certain myths about the placebo effect are worth discounting. Many think that while a placebo helps at first, it will not continue to do so. This is a misconception. In placebo-controlled trials, the effectiveness of placebo arm does not deteriorate more rapidly than that of the medication arm.

Another myth is that if a patient responds to a placebo, then his suffering some-how was not "real", but rather "all in his head" or merely being feigned. It is surpris-ing, and a bit disheartening, that physicians can still think that way, after 70 years of placebo research. If anything, the opposite will often be true: the true patient will desperately seek relief and want to believe that the remedy is helping, while the person with no actual basis may have all sorts of secondary gains which will not allow the placebo – or actual medication – to help.

Also, we should not think about placebo effects as fixed. On the individual level, they can be manipulated in endless ways, the common denominator being that the patient's expectation of recovery is enhanced. On the group level, the past 50 years have shown a diminishing gap between medication and placebo results in controlled trials for different psychiatric disorders, which reflects a growing response to pla-cebo treatments rather than a reduction in the efficacy of medication. Most of this is probably a result of unduly broadened diagnostic categories which leads to a water-ing down of the severity of the distress of the research subjects entering the proto-col; some may also be a reflection of the "professional patient" drifting from study to study in industry-supported research facilities. An additional possibility is that as psychiatric medication has become more firmly ensconced in the public consciousness, the expectation of the research subjects, and hence their placebo response, have increased.

Insofar as placebos involve expectation and conditioning, the placebo effect is hardly limited to actual placebos. Real medication will also provide benefit, at least partially, via the psychosocial effects it will have upon the patient. This is borne out by varied research findings, not only in psychiatry:

1. Medication administered without knowledge of the patient will be less effective.
2. When a new brand enters the market accompanied by marketing hype and hopeful practitioners, it will be more effective than the older brand at first, until its own effectiveness gives way to the next new medication. (We have all met the patient, desperate for help, who asks the doctor, "do you have something new to give me?")
3. Subjects receiving a more expensive brand name will respond better than those receiving a generic, even when they are all actually receiving placebo. (This is not meant to condone the high prices of many prescription drugs).

The study of the placebo is one of the fascinating and burgeoning contemporary fields in medicine, and in particular in psychiatry. But what is the psychiatrist to do with this information? Prescribing an actual placebo to a patient is clinically unwise for the rupture of trust which can ensue when discovered, unethical for the deceit involved, and potentially could expose the treating physician to medicolegal repercussions. The open label placebo research alluded to earlier solves all these problems, but despite their value, no one expects physicians to start writing prescriptions for sugar pills.

The psychiatrist who understands the potential power of the placebo effect realizes that even when the patient responds to a certain treatment, it does not necessarily mean that it was the pharmacological or neurophysiological intervention that produced the healing changes. Many clinicians, for example, would agree with the statement that "yes, the research is all very impressive, but until you're actually having a patient with severe depression who then gets a new medicine and responds, you don't understand the clinical validity of our treatments." But the sobering lesson of the research we have discussed is that the benefit does not necessarily derive from our choice of intervention, but possibly from other factors: the natural progression of the disorder (not strictly a placebo effect), the illegible Latinate lettering scrawled on a prescription pad (which we speculate might be more effective than a computer printout, but that has not to our knowledge been researched), the reassurance provided by the physician (which must be an essential component of the service we provide), and so on. It is humbling when even our therapeutic successes require us to wonder what really helped. And we remain mindful of those instances in our experience where before we had a chance to start the treatment, we were sure was necessary – say, ECT for a particularly recalcitrant depression – the person went ahead and got better, without bothering to wait for us to implement the treatment we thought so essential.

Another lesson is that we should not look down upon various other treatments. When we want to help our patients, treatment and medication ought not to be considered synonyms. For one help may come in the form of physical activity (a good first line choice for mild or moderate depression), for another a nutraceutical of the sort described later in this chapter, a third may do well on a sub-therapeutic dose, and many will be recommended a course of psychotherapy.

The profoundest lesson we draw from the placebo effect is the reminder of how complex we are, in health and in illness. We are not (for example) cars brought to the mechanic for a repair. Were we asked to enroll our vehicle in a placebo-controlled trial where half the cars received placebo engine oil, we would refrain from providing consent; and were we to read that the results of such an experiment showed an impressive placebo response, we would discontinue our subscription to that journal. But unlike cars, in addition to the physicality of our body and brain, we are also creatures who make meaning and respond to the environment and are deeply permeable to outside stimuli. Dispensing a pill will never be more than one of the many functions we perform to help our patients. More pointedly, dispensing a pill will itself always be replete with so much more than merely pharmacological significance.

The Use of Nutraceuticals in Clinical Psychopharmacology.

Nutraceuticals are a name for a broad class of food supplements or medicines that are components of normally ingested food at least in small doses but which may have effects on health in general or mental health in particular at higher than usual doses.

Neurotransmitter Precursors

Many of the common neurotransmitters such as dopamine, serotonin and norepinephrine are biochemical derivatives of amino acids found in food. A simplistic argument to raise their levels in the brain would be to give these precursors in larger amounts as drugs. Tryptophan and tyrosine are indeed sold with this claim in many health food stores. However, these amino acids are transformed into neurotransmitters by enzymes such as tyrosine hydroxylase or tryptophan hydroxylase which are rate limiting, regulated by brain molecular transcription control mechanisms and are usually saturated by the substrate tyrosine or tryptophan at normal dietary intake. L-dopa of course is an example of the opposite where the degenerating dopaminergic neurons in Parkinson's disease can have their function augmented by administration of relatively high doses of L-dopa, which is further along the metabolic pathway than dietary tyrosine and thus bypasses the block at the rate-limiting tyrosine hydroxylase enzyme. While worldwide sales of tyrosine and tryptophan as sleep aids or for mood disorders remain high, scientific evidence that they are better than placebo in any DSM-V diagnostic group is not available.

Omega – 3 Fatty Acids

Essential fatty acids are those fatty acids which are not synthesized by humans or primates and must be obtained from diet where they are most plentifully found in fish and in some plants. The cell membrane is composed to a large extent of these fatty acids' especially DHA (docosahexaenoic acid). Amazingly, MRI studies of the brain after dietary manipulation of DHA in the diet, in animals and also in humans shows that DHA composition of brain can change markedly depending on oral fatty acid quantities in the diet. Neurochemical studies have shown that the amount of fatty acids in the cell membrane affects the ability of neurotransmitter cell receptors to respond to neurotransmitter stimulation. Thus, there is biological plausibility that changes of quantity of fish consumed or administration of omega fatty acids in food supplements could affect brain function. A recent review of the literature in depression [10] found positive effects of omega-3 and particularly EPA (eicosapentaenoic acid) in depression. Positive results in depression have also been reported for EPA as an add-on to monoamine reuptake blocker antidepressant treatment [11] and even more interestingly in childhood depression [12]. However, many studies using strict DSM-V criteria for depression have not found efficacy for omega-3 fatty acids in depression. Clearly these food supplements have less pharmaceutical company backing to support large scale controlled trials than synthetic patented compounds. However, the omega-3 fatty acids have almost no side effects and almost no toxicity at overdose. Therein lies their clear advantage in the clinical practice of psychopharmacology: they can take advantage of the very, very large niche of patients for whom the clinician feels that a placebo might be indicated. Patients with mild depressions, with chronic depressions, with depression secondary to adverse social events, all of these are poor responders to monoamine reuptake blocker antidepressants and the ratio of possible benefit to possible side effects is not large. On the other hand, ethical and legal implications of giving a sugar pill labeled in the chart as placebo are problematic for clinicians in many legal jurisdictions. Prescribing or recommending without prescription an omega-3 fatty acid for these mild depressions gives the clinician an opportunity to see whether a few weeks or months might elapse with an improvement in clinical status before taking a decision as to whether a monoamine blocker might be necessary. He should also note in the chart a reference to possible proofs of clinical utility of these compounds, to avoid any charge of using placebo if he is criticized on ethical or legal grounds.

Inositol

Inositol is a simple isomer of glucose that is essential in the second messenger neurotransmitter system called the phosphytidyl-inositol cycle. It is present in many foods and is considered GRAS (generally recognized as safe). It gets into the brain poorly but high doses have been shown to enter the brain and have behavioral effects in animal models [13]. Therapeutic benefits have been reported in depression [14], in panic disorder [15] and in OCD [16]. Other studies using DSM-V criteria have not replicated these effects. Inositol, available as a powder in health food supply stores, can also like omega-3 be used as a placebo that might have biological activity but which is useful for the clinician even if it does not because of its high safety profile. The clinical charts should always quote a positive study if is recommended as treatment.

Folic Acid

Folic acid is accepted in medicine as powerful prophylactic medication for the prevention of neurodevelopmental disorders in utero and pregnant women worldwide take this vitamin from the first possibility of pregnancy or in many countries when pregnancy is even contemplated. Many studies suggest that folic acid amplifies lithium response in depression. Recently studies consistently show that homocysteine, a neurotoxic amino acid not used in protein synthesis but present in the blood, is elevated in many patients with serious mental illness [17]. Folic acid powerfully reduces serum homocysteine levels which would provide one possible biological plausibility for a therapeutic effect. A controlled cross-over study reported that folic acid reduction of homocysteine levels in chronic schizophrenia patients reduced both positive and negative symptoms [18]. A patented derivative of folic acid, methyl folate, has been approved by the FDA for depression. However, it has not been clearly shown to be superior to the simpler and widely available folic acid which can be used as above (see inositol and omega-3 fatty acids) as a safe placebo as well as a potentially therapeutic nutraceutical.

Niacin

The vitamin niacin or its derivatives was claimed by leading figures in psychopharmacology in the 1950s such as Abraham Hoffer and the Nobel Prize winner Linus Pauling to be an "orthomolecular" treatment of schizophrenia with many dramatic total cures in individual case reports. The major independent American Psychiatric Association sponsored study of niacin in schizophrenia, led by Thomas Ban was completely negative. The orthomolecular movement grew into a dogmatic cult and alienated most psychopharmacologists. That makes it less useful as an active placebo that can be incorporated into an empirical "whatever works" practice. On the other hand, patients who arrive from an orthomolecular practice already being treated with high dose niacin and who have confidence in this treatment are best not aggressively disabused of their belief system. A gentle positive acceptance is not out of place as there may be occasional patients who do indeed respond to niacin treatment as Linus Pauling and Hoffer claimed.

References

1. Berthelot JM, Le Goff B, Maugars Y. The Hawthorne effect: stronger than the placebo effect? Joint Bone Spine. 2011;78(4):335–6.
2. Hróbjartsson A, Gøtzsche PC. Is the placebo powerless? An analysis of clinical trials comparing placebo with no treatment. N Engl J Med. 2001;344(21):1594–602.
3. Khan A, Faucett J, Lichtenberg P, Kirsch I, Brown WA. A systematic review of comparative efficacy of treatments and controls for depression. PLoS One. 2012;7(7):e41778.
4. Nitzan U, Carmeli G, Chalamish Y, Braw Y, Kirsch I, Shefet D, et al. Open-label placebo for the treatment of unipolar depression: results from a randomized controlled trial. J Affect Disord. 2020;276:707–10.
5. von Wernsdorff M, Loef M, Tuschen-Caffier B, Schmidt S. Effects of open-label placebos in clinical trials: a systematic review and meta-analysis. Sci Rep. 2021;11(1):3855.
6. Amanzio M, Benedetti F. Neuropharmacological dissection of placebo analgesia: expectation-activated opioid systems versus conditioning-activated specific subsystems. J Neurosci. 1999;19(1):484–94.
7. Peciña M, Zubieta JK. Molecular mechanisms of placebo responses in humans. Mol Psychiatry. 2015;20(4):416–23.
8. Moerman D. Meaning, medicine and the 'Placebo Effect'. UK: Cambridge University Press; 2002.
9. McKay KM, Imel ZE, Wampold BE. Psychiatrist effects in the psychopharmacological treatment of depression. J Affect Disord. 2006;92(2–3):287–90.
10. Guu TW, Mischoulon D, Sarris J, Hibbeln J, McNamara RK, Hamazaki K, et al. International Society for Nutritional Psychiatry Research Practice Guidelines for Omega-3 fatty acids in the treatment of major depressive disorder. Psychother Psychosom. 2019;88(5):263–73.
11. Nemets B, Stahl Z, Belmaker RH. Addition of omega-3 fatty acid to maintenance medication treatment for recurrent unipolar depressive disorder. Am J Psychiatry. 2002;159(3):477–9.
12. Nemets H, Nemets B, Apter A, Bracha Z, Belmaker RH. Omega-3 treatment of childhood depression: a controlled, double-blind pilot study. Am J Psychiatry. 2006;163(6):1098–100.
13. Einat H, Belmaker RH. The effects of inositol treatment in animal models of psychiatric disorders. J Affect Disord. 2001;62(1–2):113–21.
14. Levine J, Barak Y, Gonzalves M, Szor H, Elizur A, Kofman O, et al. Double-blind, controlled trial of inositol treatment of depression. Am J Psychiatry. 1995;152(5):792–4.
15. Benjamin J, Levine J, Fux M, Aviv A, Levy D, Belmaker RH. Double-blind, placebo-controlled, crossover trial of inositol treatment for panic disorder. Am J Psychiatry. 1995;152(7):1084–6.
16. Fux M, Levine J, Aviv A, Belmaker RH. Inositol treatment of obsessive-compulsive disorder. Am J Psychiatry. 1996;153(9):1219–21.
17. Levine J, Stahl Z, Sela BA, Gavendo S, Ruderman V, Belmaker RH. Elevated homocysteine levels in young male patients with schizophrenia. Am J Psychiatry. 2002;159(10):1790–2.
18. Levine J, Stahl Z, Sela BA, Ruderman V, Shumaico O, Babushkin I, et al. Homocysteine-reducing strategies improve symptoms in chronic schizophrenic patients with hyperhomocysteinemia. Biol Psychiatry. 2006;60(3):265–9.

Chapter 18
Sexual Psychopharmacology: Important Symptomatic Treatment in Depression and Anxiety?

Drugs that affect sexual function are an important part of psychopharmacology, even if the drugs affect the sexual organs directly rather than through action on the brain. Some psychiatrists find themselves uncomfortable using medicines to treat erectile dysfunction, for instance, but the demands of clinical practice do not permit separating medical specialties on the basis of anatomic location of pharmacological effects. Erectile dysfunction (ED) and disorders of sexual desire and function in the male and female are common features of depression and anxiety and other psychiatric disorders. They are perhaps the most common side effects of dopamine D-2 blocking drugs for the treatment of psychosis or SSRI drugs for the treatment of depression, anxiety and OCD. The ability of the psychiatrist to flexibly use pharmacological treatments of ED and other treatments for sexual disorders without excessive dependence on urological or gynecological consultation is an important part of effective clinical psychiatry practice and enhances the doctor-patient relationship [1].

Erectile Dysfunction

The most widespread treatments of ED today are the PDE-5 inhibitors (phosphodiesterase 5 inhibitors) which were discovered by chance in a Pfizer study looking for drugs that might relax and enlarge blood vessels in the heart for the treatment of atherosclerotic angina pectoris [2]. The perspicacious investigators heard from patients about the side effect on ED and the rest is history. Phosphodiesterase is a very widely distributed enzyme in the human body that breaks down cyclic AMP or cyclic GMP, two major second messengers that are products of activation by several neurotransmitters including NO (nitric oxide) in the penile vasculature. The different isoforms of phosphodiesterase have some specificity to different areas of the body and, fortuitously, can be inhibited somewhat specifically by different chemical

R. H. Belmaker, P. Lichtenberg, *Psychopharmacology Reconsidered*, https://doi.org/10.1007/978-3-031-40371-2_18

synthetic inhibitors. The 5 isoform (PDE-5) is located in the blood vessels of the corpus cavernous of the penis and also in the female clitoral erectile tissue. It is also located in some other areas of the body including the eye which will be discussed below with relation to side effects [3].

The first PDE-5 inhibitor that reached marketing was sildenafil whose pharmacokinetics allow administration 1–2 h before intercourse and whose blood levels remain elevated for eight to 12 h afterwards. Ingestion of fatty foods simultaneously reduces the peak blood levels and total absorption fraction and so sildenafil should be taken not less than an hour after food. Doses enhance the rigidity and duration of the male erection unrelated to etiology. This fact is enormously important for the clinician to grasp and resulted almost single handedly in reversing years of theoretical writings about the importance of the distinction between psychogenic and neurogenic ED. Young men with performance-anxiety induced ED can benefit tremendously from sildenafil. Similarly, middle-aged and older men with diabetes and neuropathic damage to cavernosal nerve fibers can similarly and almost equally benefit from sildenafil treatment. The era when most men with ED were sent for nocturnal penile tumescence measurements to determine the presence or absence of REM sleep erections ground to an almost complete halt.

An equally important result of the discovery of sildenafil was the fact that sildenafil administration has no effect in the absence of sexual stimulation. See Fig. 18.1. Sexual stimulation, either by direct touch of the sexual skin or via mental fantasy, results in a highly enhanced erectile response in the presence of sildenafil. However, sildenafil alone in the absence of sexual stimulation has almost no effect, even when careful measures are made of penile tumescence. Thus, sildenafil specifically and the PDE-5 inhibitors as a class are an amazing example of the connection between mind and body: Mental stimulation's effects as exemplified by the effect of a thought or a touch on the size or shape of a whole organ can be enhanced by a specific molecule whose mechanism we understand at the chemical and physiological level.

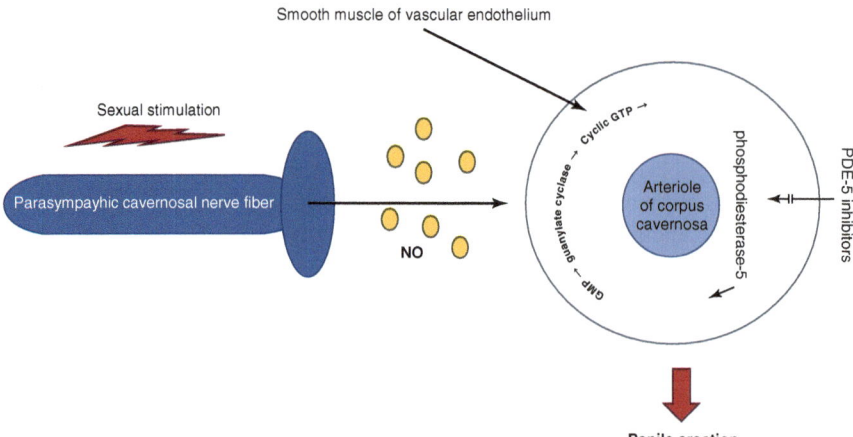

Fig. 18.1 The effect of PDE-5 inhibitors to cause penile erection after sexual stimulation

Following the tremendous commercial success of sildenafil several other conjoiners have come to market. Tadalafil is characterized by a long half-life and can even be used daily in a prophylactic manner allowing sexual activity to occur without the prior planning of taking a sildenafil tablet 1 or 2 h in advance. Vardenafil has the advantage of more rapid absorption and peak pharmacodynamic effect compared to sildenafil. Other compounds in this class are being marketed but none seems to have a significant advantage [4].

Side Effects

A rare side effect of sildenafil use is changes in color vision, perhaps due to the presence of PDE-5 in the retina. This is a rare side effect and completely reversible but is often asked about by anxious patients. No ophthalmological consultation is needed.

A common side effect of all PDE-5 inhibitors is dyspepsia due to relaxation of the smooth muscle cells in the lower esophageal sphincter. This can easily be treated with 20 mg of famotidine (an over-the-counter histamine 2 receptor blocker) simultaneously with the PDE-5 blocker or calcium carbonate antacid as necessary after PDE-5 use.

Minor side effects of the PDE-5 blockers include a red flush of vasodilation in the face and chest or mild abdominal pain and headache, none of which require treatment. In patients taking nitrates for angina pectoris, PDE-5 inhibitors for erectile dysfunction are contraindicated because of danger of severe hypotension. Patients on antihypertensive medication may experience some hypotension after use of PDE-5 inhibitors because of the synergy in relaxing smooth muscle tissue; this can be handled by reducing the dose of the anti-hypertensive medication or the PDE-5 inhibitor.

Clinical Use

Classically, male patients presenting with depression and ED were told that their sexual difficulties will improve after treatment of their depression. This is certainly sometimes true. However, a more flexible approach is indicated, depending on the clinician's evaluation of the degree of distress from the sexual dysfunction and the resulting questions about the direction of causality. Often a patient with depression and ED will improve with sildenafil treatment only and it may become clear that his depression was secondary to his erectile dysfunction. Even more common is the case of the anxious young man who suffers from ED and whose anxiety might possibly be relieved by treating the ED directly with sildenafil rather than treating the anxiety with benzodiazepines or antidepressants first. Numerous patients treated with antipsychotic drugs have elevated prolactin levels and resultant ED. Addition of a PDE-5 inhibitor to their treatment combined with a nonjudgmental, sex friendly clinician attitude can significantly improve their quality of life in many cases.

Sildenafil Abuse

Sildenafil and other PDE-5 inhibitors are not abused in the sense that patients increase their dose over time to achieve a detrimental and unprescribed effect. However, there are young men without erectile dysfunction who use sildenafil to allow themselves multiple partner erections at parties and this could have detrimental social effects. There is some black market in sildenafil perhaps for this reason or perhaps because of individuals reluctant to consult a physician [5].

Rapid Ejaculation (PE)

The ejaculatory system in the male is highly complex pharmacologically and is affected by serotonergic, alpha-adrenergic and cholinergic inputs. A powerful clinical effect to lengthen latency to ejaculation (also called the intravaginal ejaculation latency time or IVELT) is caused by SSRIs. This effect is seen in men treated for anxiety or depression with SSRIs and the increase in IVELT can become an unwanted nuisance if undesired by the female partner and unnecessary for the male partner's pleasure. In such circumstances it can be called delayed or retarded ejaculation and is listed as a side effect. However, the same effect when induced in a man suffering from rapid ejaculation can be an exhilarating therapeutic benefit. It might be hard for someone from cultures not sensitive to these issues to understand the huge suffering of a large portion of young males today who feel that they ejaculate too quickly to satisfy their female partners. This is of course a relative matter and it is hard to define rapid ejaculation outside of a dyadic relationship. Some men unrealistically judge their own ejaculatory performance in comparison with unrealistic models that they see in pornography. However, after discounting cultural and self confidence issues, rapid ejaculation is a cause of considerable objective distress. All of the SSRIs are effective, perhaps paroxetine being the most effective. At some point it was felt that it must be taken daily. However, the clinical consensus today is that a single 20 mg dose of paroxetine taken about 3 h before intercourse is usually quite effective. A specific SSRI with a very short half-life called dapoxetine was developed for marketing specifically for rapid ejaculation but is not yet approved in the USA (although it is sold in other countries}. Post marketing evaluations suggest that its efficacy is not superior enough to paroxetine to justify its increased cost as specific niche compound . The clinician therefore must explain to the rapid ejaculation patient that a drug labeled as an antidepressant is being prescribed but for another purpose, that is, rapid ejaculation. Urological work up is not necessary for a rapid ejaculation patient and the splitting of care for an anxious young man suffering rapid ejaculation into a psychiatric physician who does psychotherapy and a urologist who prescribes paroxetine is a potential danger to the clinician patient relationship. Behavioral treatments of rapid ejaculation that were once widely written about and widely taught, are usually much less cost-effective than pharmacological treatments for this disorder. Psychodynamic psychotherapy of rapid ejaculation is almost always unnecessary, ineffective and too late to save the relevant relationships [6]. Occasionally PDE-5 inhibitors can be used to help rapid ejaculation because they decrease the refractory period to achieve a second erection in young men: the second erection almost always has a much longer intravaginal latency to ejaculation and thus can be used the satisfy the partner and enhance the patient's self esteem [7].

Women and PDE-5 Inhibiters and SSRIs

The presence of phosphodiesterase isoform 5 in the clitoral erectile tissue suggested that sildenafil and similar drugs might be useful in disorders of female sexual function but has not led to a marketing indication. Some women with problems in reaching orgasm, achieving vaginal lubrication or enjoying sexual contact have reported that PDE-5 inhibitors can be helpful. Nihilism in this gender is not justified despite the failure of FDA level evidence or approved use. SSRI treatment of depression in women or anxiety in women very frequently leads to delayed orgasm as a side effect. This is almost never seen as a positive effect in the female. Occasionally PDE-5 inhibitors, by enhancing clitoral engorgement, can help reverse these SSRI side effects in the female [8].

Testosterone for Male and Female Sexual Dysfunction

Testosterone preparations are becoming increasingly available and absorption increasingly effective with the latest formulations. In men with hypogonadism, testosterone by monthly injection or by daily gel or 3 day skin patch improves both the sense of well being and sexual function. The theoretical danger of exacerbating benign prostatic hypertrophy (BPH) or increasing the rate of prostatic cancer is unsupported in recent studies although the danger is so serious that clinicians should not yet ignore it. The key problem is the definition of hypogonadism. Free testosterone in the blood is not an accurate measure since the hormone is secreted in a pulsatile manner and is mostly bound to protein and it is unclear what the appropriate laboratory cut off should be.

In females testosterone unquestionably increases sexual drive, increases number of orgasms a month and decreases time to orgasm. However, the side effects of hirsutism or deepening of voice are intolerable in most cases [9]. An FDA approved drug flibanserin for female hypoactive sexual desire disorder has a very small effect size and is rarely of clinically relevant use [10] .

HRT

Estrogen replacement for the post-menopausal woman is one of the most highly studied, written about and talked about subjects in medicine today. A psychiatrist should be fully versed in this topic and open to discussing it with all of his female patients. Menopause is a huge event for many women and the resulting decline in estrogen can cause decreased self-confidence, increased anxiety, decreased mood and decreased physical energy. Many estrogen replacement compounds are available today, sometimes combined with progesterone and sometimes without. The main danger with most of these treatments is a very slight increase in the very serious adverse effect of breast cancer which affects one of 11 women in any case. So even a very slight increase in this serious adverse effect has caused many regulatory and gynecological groups to stop recommending hormone replacement treatment as a routine matter for post-menopausal women. However, this trade-off must be evaluated individually for each woman. The psychiatrist must be closely involved and sometime is the primary care physician who may prescribe HRT after a normal gynecological examination by a gynecological specialist. HRT has been reported to be antidepressant on its own, to enhance responses to SSRI antidepressants and to improve mood, libido, energy and self-confidence in women with distress that does not meet specific psychiatric diagnostic criteria. Very low dose estrogen-like HRT is available for clinical use today and may be without danger of increased breast cancer risk although the topic has not been fully studied [11].

Transgender Treatment

Transgender treatment involves a psychiatrist in diagnosis and often in the treatment of gender dysphoria at some point for most patients. The endocrinological treatment of gender transition is highly specialized and must be handled by an endocrinologist.

Sexology, Psychopharmacology and Psychiatry

It is hard to imagine using psychotropic drugs without first becoming aware of the symptoms, distress and behaviors which they treat. It is equally difficult to imagine treating sexual disorders without the physician becoming comfortable discussing these symptoms and aspects of human behavior. Sexology is an important part of psychiatry and sexual psychopharmacology is an important part of psychopharmacology [12].

Clinical Vignettes

1. Adam was a 25 year old shy man who had been involved for 3 months in a wonderful first love, but simultaneously developed an overwhelming generalized anxiety disorder that brought him to a psychiatrist. He and his partner were having sex twice weekly, and he ejaculated rapidly each time. His partner told him that this was fine, but indirectly let him know that previous boyfriends had lasted longer and that she was not having orgasms. Paroxetine 20 mg 2 h before sex for him led to first orgasms with him for her. His anxiety disappeared.
2. A 60 year old married college professor had been in the doldrums for 2 years. There was no diagnosis. He and his wife enjoyed sex and many other things, but she let slip that she orgasmed with her vibrator in their couple play but she missed his youthful vigor, trying to be very general in her meaning. Sildenafil 50 mg 1 h before intercourse gave him the firm, longer erections that she remembered and loved, and their general enjoyment of life improved for both of them.
3. Ruth was 49 and was referred after many general bodily complaints were found to be of no organic cause, and her family physician had not gotten a response to lorazepam or fluoxetine. She had stopped menstruating at age 44. Eviana (estradiol 0.5 mg + norethisterone 0.1 mg) daily led to an increase in general sense of wellbeing and total decrease of somatic complaints over 3 months.

References

1. Shamloul R, Ghanem H. Erectile dysfunction. Lancet. 2013;381(9861):153–65.
2. Goldstein I, Burnett AL, Rosen RC, Park PW, Stecher VJ. The serendipitous story of sildenafil: an unexpected Oral therapy for erectile dysfunction. Sex Med Rev. 2019;7(1):115–28.
3. Etminan M, Sodhi M, Mikelberg FS, Maberley D. Risk of ocular adverse events associated with use of phosphodiesterase 5 inhibitors in men in the US. JAMA Ophthalmol. 2022;140(5):480–4.
4. Mykoniatis I, Pyrgidis N, Sokolakis I, Ouranidis A, Sountoulides P, Haidich AB, et al. Assessment of combination therapies vs monotherapy for erectile dysfunction: a systematic review and meta-analysis. JAMA Netw Open. 2021;4(2):e2036337.
5. Kamin R, Zion IB, Chudakov B, Belmaker RH. Sildenafil effects on sexual function in asymptomatic volunteers: a controlled study. J Sex Marital Ther. 2006;32(1):37–42.
6. Pryor JL, Althof SE, Steidle C, Rosen RC, Hellstrom WJ, Shabsigh R, et al. Efficacy and tolerability of dapoxetine in treatment of premature ejaculation: an integrated analysis of two double-blind, randomised controlled trials. Lancet. 2006;368(9539):929–37.
7. Krishnappa P, Fernandez-Pascual E, Carballido J, Martinez-Salamanca JI. Sildenafil/Viagra in the treatment of premature ejaculation. Int J Impot Res. 2019;31(2):65–70.

8. Nurnberg HG, Hensley PL, Heiman JR, Croft HA, Debattista C, Paine S. Sildenafil treatment of women with antidepressant-associated sexual dysfunction: a randomized controlled trial. JAMA. 2008;300(4):395–404.
9. Chudakov B, Ben Zion IZ, Belmaker RH. Transdermal testosterone gel prn application for hypoactive sexual desire disorder in premenopausal women: a controlled pilot study of the effects on the Arizona sexual experiences scale for females and sexual function questionnaire. J Sex Med. 2007;4(1):204–8.
10. Aftab A, Chen C, McBride J. Flibanserin and its discontents. Arch Womens Ment Health. 2017;20(2):243–7.
11. Lobo RA, Gompel A. Management of menopause: a view towards prevention. Lancet Diabetes Endocrinol. 2022;10(6):457–70.
12. McVary KT. Clinical practice. Erectile dysfunction. N Engl J Med. 2007;357(24):2472–81.

Chapter 19
Neuropsychopharmacology: Huntington's, Parkinson's, Tardive Dyskinesia, Narcolepsy, Migraine, Tourette's and Schizophrenia-Like Psychosis of Epilepsy

The borders between psychiatry and neurology are blurry and change over time, with the training of particular physicians, and with the development of new treatments. Several illnesses of the brain have marked behavioral symptomatology although they also have clear neurological pathology and are often treated by neurologists. Psychopharmacologists are often necessary in consultation and sometimes even become the primary physician. One very common such disorder, Alzheimer's has already been discussed in Chap. 13. Seven less common disorders will be discussed here both because they are important in common psychopharmacological practice and they are instructive regarding the mechanisms of action of psychotropic drugs.

Huntington's

Huntington's disease is one of the more common single gene disorders and is inherited in a dominant form [1]. The age of onset is often in the 30 s or 40 s of life after reproduction has taken place and so a family history is often enough to make the diagnosis. The exact gene that is mutated in this condition has been identified and its abnormal protein product in the brain defined but no rational therapy has yet been derived from the molecular genetic discovery. The patient usually presents with abnormal choreiform movements of the face or limbs that immediately remind any psychiatrist of tardive dyskinesia. Indeed, these movements are easily treated with small doses of dopamineD-2 receptor blocking first generation antipsychotics. The antipsychotics are effective in Huntington's patients for years or decades until they develop a more generalized dementia unresponsive to dopamine blocking therapy. By mistake, Huntington's patients are sometimes diagnosed as Parkinson's and given anticholinergic drugs. This worsens their abnormal movements and if a diagnosis was not made initially, it will be obvious to the clinician when this worsening

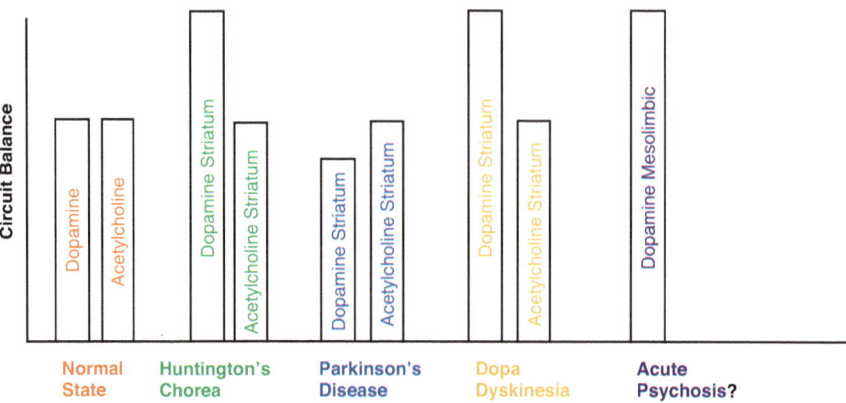

Fig. 19.1 Neuropsychopharmacology involves balances between various neurotransmitters

occurs. Fig. 19.1 illustrates the theoretical balance of the cholinergic and dopaminergic circuits in Huntington's vs. psychosis vs tardive dyskinesia. In the basal ganglia, deep areas of the brain that have controlled posture and movement since the evolution of the early vertebrates, cholinergic and dopaminergic systems are in balance. In Huntington's it is postulated that a deficit in GABAergic inhibition of the dopaminergic system leads to dopamine hyperactivity relative to cholinergic activity. Anti-dopamine treatment restores the balance. Anticholinergic treatment worsens the imbalance. Huntington's patients often are in need of psychiatric counseling for depression and anxiety related to genetic counselling for their children or grandchildren, their guilt feelings for passing on this gene and their existential concern about their future course which is usually well defined deterioration [1, 2].

Narcolepsy

Narcolepsy is a sleep disorder characterized by sudden onset of sleep during the daytime period without an appropriate setting and despite full nighttime sleep. Polysomnography shows that the sleep attacks are characterized by rapid transition into REM sleep [3]. Just as in REM sleep there is muscle cataplexy, sometimes these patients also have attacks of total body loss of motor tone in inappropriate settings called cataplexy. This disorder has long been treated with classical antidepressants such as impramine with good results or with stimulants such as amphetamine with mixed results. Modanifil and its congeners are also dopaminergic stimulants with unclear mechanisms that are not clearly more effective than amphetamine but have less abuse potential [2]. In recent years the pathophysiology of the disorder has become clear: It is a neurodegeneration of hypocretin containing neurons localized in the hypothalamus. The neurodegeneration is autoimmune in origin. Often these patients require a psychiatrist to evaluate their suitability for work potential or their need for social security disability and the psychiatrist should of course be aware of the nature of their disease as well as psychosocial effects on exacerbation or amelioration of the syndrome [4].

Parkinson's

Parkinson's disease is perhaps the most familiar neurological disease for most psychiatrists, at least in previous generations because first generation antipsychotic drugs had such prominent parkinsonian side effects [5]. Parkinsonian side effects in first generation antipsychotic drugs were always treated with anticholinergic drugs because it was known early on that L-dopa treatment worsened psychosis.

Parkinson's disease is one of the few neuropsychiatric illnesses where treatment was developed on a rational basis. Parkinson's is a common disease with an incidence similar to schizophrenia of about one percent of the population. However, the prevalence is lower than schizophrenia since the age of onset is usually over 60 years old rather than in the 20 s as in schizophrenia. Thus, even though one percent of the population may be affected over their lifetime, fewer Parkinson's disease patients live in the population at any given time compared to the number of schizophrenia patients. Nevertheless, Parkinson's causes considerable disability and mortality. The symptoms include the classic triad of tremor, rigidity and akinesia. The parkinsonian tremor has a classic frequency and body distribution, and is often described as the act of fingers counting money bills. The akinesia when it affects the facial muscles can lead to a blank stare or dulling of the outward expression of emotion that can be confused with the flatness of affect in schizophrenia or the depressed affect in melancholia. The gait disturbance involving small steps that propel the patient forward in a way that makes him seem as if he is about to fall over can be so diagnostic that the physician can raise this diagnosis as the patient walks into the office. The Parkinson's disease that was first described eponymically was found in the 1950s to be due to the deterioration of the dopamine containing neurons in the substantia nigra. On that basis L-dopa, the precursor of dopamine biochemically, was given as a potential drug treatment. The story of that discovery is a story of heroic clinical therapeutics. Dopamine itself could not be given because it does not cross the blood brain barrier. Low doses of L-dopa were not helpful and were reported by many investigators as non-therapeutic. Only by raising the dose above anyone else's predications did Cotzias in the 1960's find the miraculous therapeutic effect of L-dopa in Parkinson's [6]. This has become the mainstay of treatment. Interestingly, L-dopa is effective only as long as a minimum number of dopaminergic neurons remain alive in the substantia nigra. These neurons can take up and use the externally administered L-dopa to convert it into dopamine and restore synaptic neurotransmission (the paradoxes and mechanisms involved in this effect are wonderous but too complex to be discussed here). However, when the final dopamine cells die L-dopa is no longer effective. Anticholinergic treatment is also effective in Parkinson's Disease, as it is in antipsychotic induced parkinsonism. See fig. 18–1 as to how the lowered dopamine vs. acetylcholine balance in Parkinson's can be restored by cholinergic blockers. However, the cholinergic blockers impair memory (see Chap.13) and neurologists are hesitant to recommend them today. Monoamine oxidase inhibitors prevent the breakdown of dopamine in the dopaminergic synapse and were originally used as antidepressants (see Chap. 5). However, they are somewhat useful in Parkinson's disease and this could easily be understood because they might enhance the level of dopamine by preventing its breakdown. However, commercially motivated claims that a more recent MAO inhibitor rasagiline, also prevents the progression of neurodegeneration in Parkinson's Disease are unsupported by the majority of the clinical studies today.

Direct dopamine receptor stimulation, which bypass the deteriorating presynaptic nigrostriatal neurons, can also be effective in Parkinson's disease. Such drugs include pramipezole or ropinirol. Parkinson's patients usually develop a more generalized Alzheimer's like dementia in the later phases of the illness after 5–15 years of progression. This Parkinson's dementia is unaffected by L-dopa. Parkinson's patients present with depression much more frequently than the severity of their illness alone can explain and deterioration of monoamine neurons may extend beyond the dopamine neurons to other systems as well. That could be the basis of their depression. Antidepressants are often used successfully in such patients [7].

A common reason for cessation of L-dopa therapy is the development of L-dopa psychosis. This fact supports the dopamine theory of psychosis as discussed in Chap. 6 although there is no direct proof of this theory in non-parkinsonian psychosis. Treatment of Parkinson's psychosis with dopamine blocking antipsychotics is contraindicated because these drugs will exacerbate the underlying Parkinson's disease. Lowering the dose of L-dopa is often successful in ameliorating Parkinson's psychosis. In cases where it is not possible to find a dose of L-dopa that ameliorates the Parkinson's motor symptoms without inducing a psychosis, clozapine has been proven to be effective in ameliorating psychosis without exacerbating the Parkinson's Disease. The mechanism is not clear but many have speculated that serotoninergic blockade is involved. The FDA has approved the use of pimavanserin, a specific serotonin 5-HT2 receptor blocker, for Parkinson's Disease psychosis and this would support the interaction of serotoninergic systems for clozapine's specific mechanism of action.

Another frequent reason for cessation of L-dopa in Parkinson's disease treatment is also of interest to the psychiatrist: Parkinson's patients treated over time with L-dopa will sometimes develop choreiform movements of the limbs that are not rapid tremor-like movements but are slow movement of the whole limb or the mouth, lips and tongue. These oral buccal lingual dyskinesias or distal limb choreiform movements seem to be manifestations of dopamine receptor super sensitivity that might be due to long term denervation in the deteriorating nigrostriatal tract. The post synaptic receptors then become supersensitive to the exogenous L-dopa. The clinical symptomology is reminiscent of Huntington's and supports the overall pattern presented in fig. 18–1 illustrating the relationship between Parkinson's Disease, Huntington's Disease, psychosis and tardive dyskinesia. However, the devil is in the details and these models must be understood as over simplifications that help us guide treatment. The neuropsychopharmacology of Huntington's and Parkinson's is clearly an inspiration by analogy for the psychopharmacologist of psychosis but psychosis in the context of schizophrenia or bipolar disorder stands out for its lack of the defined anatomic pathology that Huntington's and Parkinson's disease possess [5–7].

Tardive Dyskinesia

Tardive dyskinesia is a syndrome of involuntary choreiform motor movements of the fingers, extremities and particularly oral buccal lingual (OBL) area. The syndrome occurs most often in persons treated with first generation antipsychotic drugs. However, it also has an incidence in the population unrelated to antipsychotic drugs, particularly in elderly edentulous people. When first generation antipsychotics were the mainstay of treatment, tardive dyskinesia was an extremely serious problem. Today it is much less common but still exists. The similarity of tardive dyskinesia to Huntington's Chorea and to dopa dyskinesias is apparent to every clinician. If the presumably causative antipsychotic drug was stopped, the dyskinesia usually became worse. Anticholinergic, antiparkinsonian agents made tardive dyskinesia worse just like in Huntington's chorea. These facts suggested that tardive dyskinesia is a a dopamine receptor supersensitive syndrome caused by a cellular process compensating for chronic inhibition. Raising the dose of the antipsychotic that presumably caused the dyskinesia was found to remove or even eliminate the dyskinesia in most cases. It was shown in several studies that the dyskinesia rarely reappeared after dose increase such that the physician did not face a situation akin to tolerance or addiction. However, clinicians and the legal system were uncomfortable with the concept that tardive dyskinesia should be treated by increasing the dose of the offending agent. A slew of new treatments have been patented and are marketed today vigorously. They all involve a reserpine based mechanism that involves depleting dopamine in the presynaptic vesicles. This leads to the same dopamine function decrease that would occur if the dose of dopamine blocking agent had been increased. Moreover, reserpine itself and these agents such as tetrabenazine also have been reported to cause tardive dyskinesia. The best clinical judgment would still be to treat tardive dyskinesia by increasing the dose of dopamine receptor blocking antipsychotic, especially if the neuroleptic is still necessary for treatment of psychosis [8].

Migraine

Migraine headaches are frequently seen by psychiatrists and often fall between the cracks of their psychiatry and neurology appointments [9]. The headaches are severe, often unilateral, often begin with severe eye pain and often have an impending aura. This distinguishes them from muscle tension headaches which have less sudden onset, less defined offset, no aura and respond to muscle relaxation treatment including benzodiazepines. Migraine's pathophysiology has had many reincarnations over the last few decades but most modern theories center on the cerebral vasculature and its many sensitive pain receptors. Current treatments involve aborting an attack as soon as it is recognized with a triptan medication such as sumatriptan: these are 5-HT receptor agonists and may work at receptors in the cerebral vasculature or in the brain [10]. Prophylactic treatment is often effective with anti-epileptic drugs such as topiramate or valproic acid but also with b-adrenergic blockers including propranolol. The medications involve pharmacology familiar to psychopharmacologists but with incomplete overlap and confusing contradictions. Migraine is common in patients with anxiety disorder and depression but the reason for comorbidity is unclear. Uncontrolled migraine can often cause anxiety and depression. The psychiatrist must be aware both of the differential diagnosis of migraine headaches from tension headaches and the interaction of the treatment of migraine with any psychotropic drugs for anxiety or depression that he might prescribe [9, 10].

Tourette's Syndrome

Tourette's Syndrome first appears in childhood and is a dramatic illness of multiple tics including usually prominent vocal tics [11]. The vocal tics may be unintelligible barks but are usually curse words. These words are of course culture and language group specific but on the other hand remarkably identical in cultures all around the world. No single gene has been identified in large scale genome scanning studies. The syndrome is associated with considerable social ostracism and loss of self confidence such that much psychotherapy and family therapy is needed. At least a third of the cases also have comorbid classic obsessive compulsive disorder (OCD, see Chap. 14).

Tourette's Syndrome patients respond to classic dopamine D-2 receptor blockers with marked reduction in their motor tics including the vocalizations. While pimozide used to be the drug of choice, risperidone is more commonly used today. There is no evidence that one D-2 blocker is more effective than another. The usual side effects include Parkinsonism, increased prolactin and delayed puberty in boys may occur. Newer dopamine depleting agents, congeners of tetrabenazine, seem no more effective than dopamine D-2 blockers [12]. Pharmacotherapy at the minimum necessary dose together with psycho-education and appropriate placement can change these children and adult patients' lives for the better and sometimes lead to total normalization [13].

Schizophrenia-like Psychosis of Epilepsy

Epilepsy is a common disorder with numerous subtypes. It is treatable by a wide range of anti-epileptic medicines available to the neurologist requiring follow up dosage adjustment and often blood level monitoring. Despite care that often achieves total or almost total reduction in epileptic attacks, epileptic patients have high prevalences of anxiety and depression, more than expected in comparison with illnesses that give similar degrees of disability. The most serious psychiatric complication is the development of a psychosis that most typically includes religious and grandiose delusions. The acute psychosis almost always gradually progresses to negative symptoms that include flattened affect, loosening of associations and cognitive difficulties similar to some forms of schizophrenia occurring without epilepsy. It is difficult to find a name for this syndrome other than the classic one "schizophrenia like psychosis of epilepsy". Unfortunately, and counterintuitively, both the positive and negative symptoms of this syndrome respond poorly if at all to D-2 receptor blockers. There seems to be little relationship between success of the treatment of the epileptic disorder and the occurrence of this psychiatric syndrome. Usually, psychiatrists rather than neurologists are the ones who treat the psychosis in epileptic patients. It is important for such psychopharmacologists not to overpromise results with antipsychotic medication and to involve the family in realistic goals which may be particularly confusing and disappointing for them if they had experienced a positive and optimistic therapeutic response to their loved one's epilepsy itself. Schizophrenia-like psychosis is most likely if the epileptic focus is in the temporal lobe and/or the symptoms are complex partial seizures (temporal lobe epilepsy).

References

1. Gusella JF, Lee JM, MacDonald ME. Huntington's disease: nearly four decades of human molecular genetics. Hum Mol Genet. 2021;30(R2):R254–r63.
2. Belmaker RH. Modafinil add-on in the treatment of bipolar depression. Am J Psychiatry. 2007;164(8):1143–5.
3. Andlauer O, Moore H, Jouhier L, Drake C, Peppard PE, Han F, et al. Nocturnal rapid eye movement sleep latency for identifying patients with narcolepsy/hypocretin deficiency. JAMA Neurol. 2013;70(7):891–902.
4. Mahoney CE, Cogswell A, Koralnik IJ, Scammell TE. The neurobiological basis of narcolepsy. Nat Rev Neurosci. 2019;20(2):83–93.
5. Bloem BR, Okun MS, Klein C. Parkinson's disease. Lancet. 2021;397(10291):2284–303.
6. Verschuur CVM, Suwijn SR, Boel JA, Post B, Bloem BR, van Hilten JJ, et al. Randomized delayed-start trial of levodopa in Parkinson's disease. N Engl J Med. 2019;380(4):315–24.
7. Armstrong MJ, Okun MS. Diagnosis and treatment of Parkinson disease: a review. JAMA. 2020;323(6):548–60.
8. Factor SA, Burkhard PR, Caroff S, Friedman JH, Marras C, Tinazzi M, et al. Recent developments in drug-induced movement disorders: a mixed picture. Lancet Neurol. 2019;18(9):880–90.

 9. Eigenbrodt AK, Ashina H, Khan S, Diener HC, Mitsikostas DD, Sinclair AJ, et al. Diagnosis and management of migraine in ten steps. Nat Rev Neurol. 2021;17(8):501–14.
10. VanderPluym JH, Halker Singh RB, Urtecho M, Morrow AS, Nayfeh T, Torres Roldan VD, et al. Acute treatments for episodic migraine in adults: a systematic review and meta-analysis. JAMA. 2021;325(23):2357–69.
11. Muth CC. Tics and Tourette Syndrome Jama. 2017;317(15):1592.
12. Jankovic J, Coffey B, Claassen DO, Jimenez-Shahed J, Gertz BJ, Garofalo EA, et al. Safety and efficacy of flexible-dose Deutetrabenazine in children and adolescents with Tourette syndrome: a randomized clinical trial. JAMA Netw Open. 2021;4(10):e2128204.
13. Kurlan R. Clinical practice. Tourette's Syndrome N Engl J Med. 2010;363(24):2332–8.

Chapter 20
Medication for Eating Disorders: By Popular Demand?

Eating and weight control are essential aspects of survival for all organisms and are controlled by multiple mechanisms in the gastrointestinal tract and in the brain. The world-wide epidemic of obesity is caused by increased availability of high caloric foods, advertising of fattening foods and cultural trends in eating. A less well known but highly problematic disorder is anorexia nervosa which is also increasing in incidence in most developed countries in the world, perhaps because the obesity epidemic causes pressure on some young women to diet and lose weight. This chapter will discuss the limited pharmacological tools at our disposal today for therapy of eating disorders, although it should be clear from the start that all of these disorders require a multidisciplinary, comprehensive and primarily psychosocial approach.

Obesity

Popular demand for medical weight reducing treatment is fueled by the difficulty and lack of success of dieting for most individuals. One mechanism affecting hunger and satiation is the brain serotonergic system and especially the serotonin 5-HT2 receptors. Amphetamines release monoamines including serotonin onto many receptors including 5-HT2 receptor and this is thought to be the mechanism of the well demonstrated effect of amphetamine and its congeners to reduce appetite and induce some weight loss. Weight loss is a consistent and well documented side effect of amphetamine therapy in attention deficit disorder (see Chap. 11). Amphetamines were widely distributed in the United States and Europe a half century ago for weight loss but were universally withdrawn from medical practice for this indication because of the risk of psychosis (as discussed in Chap. 11). Derivatives of amphetamine have since then been sold and approved by the FDA such as sibutramine which reached wide sales for a brief time before being withdrawn by the FDA because of serious side effects; fenfluramine which was withdrawn for side effects;

R. H. Belmaker, P. Lichtenberg, *Psychopharmacology Reconsidered*, https://doi.org/10.1007/978-3-031-40371-2_20

and phentermine which is still available in a limited manner. No amphetamine-based treatment of obesity is effective and safe long-term.

Another approach to weight reduction has been the use of glucagon-like peptide receptor agonists. Several have been developed for diabetes control. In much higher doses than used for diabetes, they reduce appetite and successfully reduce weight in study populations. The main side effects are nausea and abdominal pain because of disordered motility of the gastrointestinal tract. It is possible that the reduced appetite is secondary to the nausea and abdominal pain and it is not yet clear if these currently popular medications will be helpful in the long term and in larger phase IV study groups. It is not clear of the glucagon-like peptide receptor agonists have any CNS effects or feedback emotional effects from gastrointestinal tract feedback neurons [1].

Anorexia Nervosa

Anorexia nervosa is an illness characterized by markedly reduced food intake and BMI (body mass index) to the point where the symptoms interfere with normal physiology and energy levels including cessation of menses. It is far more common in young women than in any other group. The weight loss is almost always accompanied by a distorted body perception where the patient sees herself as fat when she is really emaciatingly thin. This is truly what she sees when she looks at herself in the mirror and is often of a delusional quality. However, antipsychotic medications are of no help in anorexia nervosa. While cannabis increases appetite in many users (see Chap. 15) and William Osler the famed physician over a century ago proposed it for anorexia nervosa, it has not been found to be helpful. Antidepressants can be helpful if the anorexia nervosa patient develops a depression but usually are useless in the central disorder of weight loss and abnormal body perception. Unfounded optimism about the use of psychopharmacology agents in these teenagers, often who have BMIs that plummet within months to levels bordering on the necessity for hospitalization and parenteral feeding, can be dangerous. Anorexia nervosa requires rapid referral to a multidisciplinary team sometimes including day hospitalization and daily group psychotherapy and there is no psychopharmacological cure [2].

Bulimia

Bulimia is an illness characterized by binges of eating of thousands of calories, often but not always ice cream or sweets, within the course of a few hours followed by vomiting induced by the patient often by putting a finger up the throat. This illness occurs usually in young women and some studies have found prevalence of tens of percent of the population in some female student groups. Bulimia can cause serious medical complications because the vomitus contains an unbalanced composition of ions and can leave the patient with sodium, potassium and chloride imbalances in the blood. The vomitus passing though the mouth often leads to pathognomic dental and gum disorders in these patients. By contrast with anorexia nervosa, antidepressants are very effective in bulimia and often relieve the disorder entirely. The synergism of psychotherapy is often marked. Antidepressants combining both serotonin and norepinephrine reuptake blockade, such as the original imipramine or the modern venlafaxine, are most effective [3].

Conclusions

Anorexia nervosa is one of the few disorders discussed in this textbook where excessive optimism about psychopharmacology can delay appropriate treatment, sometimes with fatal consequences. The contrast in this regard between anorexia nervosa and bulimia is a humbling example of our lack of understanding.

References

1. Treasure J, Duarte TA, Schmidt U. Eating disorders. Lancet. 2020;395(10227):899–911.
2. Mitchell JE, Peterson CB. Anorexia Nervosa. N Engl J Med. 2020;382(14):1343–51.
3. Weinstein JJ. Bulimia nervosa. N Engl J Med. 2003;349(24):2363–4. author reply -4

Chapter 21
Pharmacological Treatment of Alcohol Abuse: A Short Chapter

Introduction

Alcohol is a substance that causes mild pleasure to billions around the globe but addiction, life threatening physical disease and havoc in the home and work place to many hundreds of millions. In some countries as many as half of internal medicine hospital beds are occupied by patients with chronic alcoholic liver disease, gastric disease or cardiovascular disease. The situation is analogous to that of nicotine cigarettes but efforts to prohibit all use of alcohol in the United States in the 1920s and in a few other countries were disastrous and unenforceable. There have been unfortunately few other major national level social efforts to reduce alcohol advertising and cultural approbation since then. Movies and television seem to glorify alcohol use [1].

Ethanol or ethyl alcohol is a small organic molecule that has been known and used since antiquity. Its closest pharmacological relatives in terms of its mechanism of action in the brain are the benzodiazepines that act on the GABA receptor complex allosterically to induce relaxation or sleep. However, alcohol does not act at the benzodiazepine binding site of the GABA receptor complex; instead its main action seems to be to directly open chloride channels associated with GABA receptors and thereby enhance GABA neurotransmission. At higher doses as with benzodiazepines alcohol can cause coma and respiratory depression. At low doses, as with benzodiazepines, its ability to relax involves suppression of cortical inhibitory circuits resulting in a pleasant prosocial disinhibition [2].

A significant percentage of alcohol users become addicted: Whether this is 1% or 10% of the users depends on cultural setting and reporting methodology. Attempts to treat alcohol addiction by replacement of the alcohol with benzodiazepines are not successful because the patients almost universally use the two together resulting in the danger of synergistic effects on function and respiratory depression. An old treatment of alcohol addiction was the use of disulfiram, an inhibitor of

R. H. Belmaker, P. Lichtenberg, *Psychopharmacology Reconsidered*, https://doi.org/10.1007/978-3-031-40371-2_21

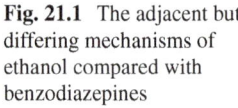

Fig. 21.1 The adjacent but differing mechanisms of ethanol compared with benzodiazepines

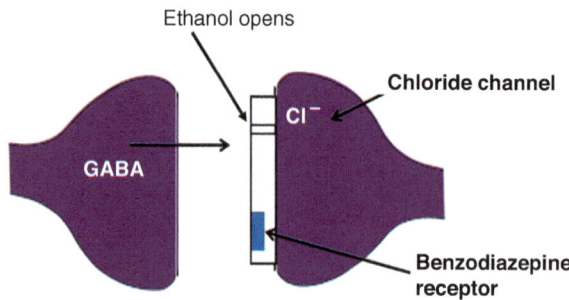

acetaldehyde dehydrogenase which is an enzyme necessary in a step of ethanol's metabolism in the blood. Disulfiram's daily administration causes a patient who uses alcohol to have a highly unpleasant vomiting response to ingestion of alcohol because of the buildup in his blood of acetaldehyde. However, most patients either stop the disulfiram when they want to go on an alcohol binge or expose themselves to the dangers of the disulfiram reaction which in some cases can cause death.

A recent pharmacological treatment of alcohol abuse is selincro (nalmefene). This is an opioid antagonist similar to naltrexone and it use is based on studies that show that alcohol abuse is partially mediated by pleasure mediated by opiate pathways (after the initial alcohol effect at the GABA receptor complex). This medication is reported in several studies to reduce desire for alcohol use in patients without alcoholism but with problematic excessive drinking. The drug does not seem to be.

Patients with alcoholism need comprehensive psychosocial care often including inpatient detoxification, close follow-up and family therapy. Withdrawal from alcohol addiction poses a serious risk of delirium tremens and seizures with potential mortality. Treatment of alcohol withdrawal or delirium tremens involves high-dose intravenous benzodiazepines, often chlordiazepoxide, sometimes for 10 days (See Chap. 7). The electrolyte disturbances and physical complications necessitate an internal medicine specialist, but most community follow-up treatment is done by psychiatrists. The equivalence of benzodiazepines for alcohol in withdrawal supports the concept (see Chap. 7) that alcohol and benzodiazepines share a molecular target of the GABA receptor (see Fig. 21.1). Future developments may include long-acting medications analogous to methadone maintenance in opiate addiction, but at present there are no specific pharmacological treatments of alcohol addiction or misuse [3].

References

1. Carvalho AF, Heilig M, Perez A, Probst C, Rehm J. Alcohol use disorders. Lancet. 2019;394(10200):781–92.
2. Fuster D, Samet JH. Alcohol use in patients with chronic liver disease. N Engl J Med. 2018;379(13):1251–61.
3. Kranzler HR, Soyka M. Diagnosis and pharmacotherapy of alcohol use disorder: a review. JAMA. 2018;320(8):815–24.

Chapter 22
Is There a Potential for New Treatments in Psychopharmacology or Have We Picked All the Low Hanging Fruit?

RHB

This volume has been about psychopharmacology today and we have not reviewed the many exciting current investigations of possible new directions for the future. Psychopharmacology today consists almost entirely of five groups of compounds: Dopamine receptor blocking antipsychotics; monoamine reuptake blocking antidepressants (also useful in anxiety); benzodiazepine receptor allosteric agonists of the GABA receptor (useful in anxiety and insomnia), mood stabilizers including lithium, anti-epileptic mood stabilizers, and second generation antipsychotics; and stimulants for childhood hyperactivity and attention disorder. These compounds were all discovered serendipitously and the muse of serendipitous discovery seems to be hibernating over the last two decades. Instead, pharmaceutical companies have churned out "me-too" compounds that may have slightly different pharmacokinetics or side effects and have been promoted vigorously as the patent expired on the previously most popular compound in its class. There have been many gloomy review articles with titles such as "Why has the pipeline in neuropsychopharmacology dried up?" It is not clear why serendipitous major discoveries in clinical neuropsychopharmacology have been in a paused state, but in contrast basic neuroscience has been experiencing exponentially increased knowledge in the last 20 years [1]. Most of it has not been relevant for the psychiatric clinician. Additionally, there has been in neurosciences and particularly in behavioral neuroscience a "replicability crisis": Numerous articles have been retracted and others replicate only for the exact animal strain in which they were first published or depend on extremely specific conditions [2]. Results that depend on such specific conditions are hard to imagine as findings that can be extrapolated from rodents all the way to human beings.

What then are the chances for future developments that <u>are</u> clinically relevant in psychopharmacology? It is as hard to predict scientific advances as it is to predict the stock market. The human brain is probably the most complex system in the

R. H. Belmaker, P. Lichtenberg, *Psychopharmacology Reconsidered*, https://doi.org/10.1007/978-3-031-40371-2_22

universe and there are hundreds or thousands of different neurotransmitters, neuro-hormones and modulatory agents. Most of these affect second messenger systems inside the cell and many then affect third messenger systems that control transcription or translation of the DNA genome promotor sites. It would be hard to believe that we will not find new drugs that affect some of these sites and that will be possible to use therapeutically in various areas of human psychiatric suffering.

How can we maximize the chances that we will find such new pharmacological treatments in the shortest possible time? Clearly using the DSM-V idea of specific diagnoses to test the effectiveness of new compounds has been a failure and could continue to hobble our efforts if it is not reworked. Of the therapeutic compounds in psychopharmacology that we have today, none (*absolutely none*) are specific to a specific DSM-V diagnosis. Moreover, it is clear that no psychiatric diagnosis is caused by a specific single gene pathway or is associated with a specific pattern on MRI imaging or EEG. We may be more likely to find therapeutic compounds in the future if we test for compounds that affect anxiety unrelated to DSM-V diagnosis; sadness or low mood unrelated to DSM-V diagnosis; psychosis unrelated to DSM-V diagnosis; or substance addiction unrelated to DSM-V diagnosis. Search for effects on these human emotions could be more easily and directly related to effects on animal behavior and neurochemical circuits in animals.

Another suggestion to increase our chances of finding new treatments would be to change the FDA demands for two placebo-controlled trials in psychiatric disorder. This FDA demand merely encourages me-too development. Better approaches could be to test new drugs as add-ons to current treatment to look for additive effects of novel new compounds or to compare new drugs head-to-head with existing compound to require superiority compared to existing treatment.

Increasing options in the future for biological treatments in psychiatry other than psychopharmacology include transcranial magnetic brain stimulation [3] or deep brain highly localized electrical stimulation with indwelling electrodes as in resistant Parkinson's Disease. A combination of these methods might be imagined such as an indwelling catheter delivering a pharmacological agent to a specific brain site at doses higher than what could be achieved via systemic administration and without the peripheral side effects that accompany systemic administration [4]. All pharmacological approaches that use the systemic administration of a drug are limited by the fact that a drug reaches all parts of the brain. Any specificity that we can achieve depends on the specificity of that drug's molecular structure and the resultant interaction with specific brain receptor sites or specific molecules involved in second messenger or third messenger signaling. Thus, advances in synthetic chemistry will be a key to future progress in psychopharmacology. The great advances in neuroscience must be translated into targets that are "druggable" which is a complex but extremely important field.

Advances in neurosciences can induce enthusiasm among psychiatrists particularly in those who are young and future oriented. However, it can be disappointing to families with ill loved ones or to practical clinicians that need an answer today. Our book is pointedly directed to the last group for whom overpromising by neuroscientists has sometimes been discouraging. The contributions of psychopharmacology over the last half century have been remarkable and we have tools to help many patients but only a few cures and many side effects. Optimal use of the medicines we have today may be promoted by using these medications in a "today" oriented framework, separate from the hopes for neuroscience for the future. A dialectical approach of maintaining optimism for the future while treating with realism today is the note on which we end this book.

PL

As the reader will have understood, there is much to fret about in the contemporary state of psychopharmacology. Can we be more hopeful about the future? On this point, I suspect that the tone of optimism in the first part of this chapter, though qualified, is still overstated. I feel less sanguine as I look to the future. I accept that there are myriad candidates for psychopharmacological manipulation which have yet to be discovered, much less clinically evaluated. But doubt about the prospects for greater success in the future seems to me wiser. With the perspective of 70 years of psychopharmacology, we can grasp that we have indeed picked the low-hanging fruit, but there is no reason to expect that the less accessible fruit will be tastier. Our experience till now should be enough to sober our expectations.

One lesson that our hitherto limited psychopharmacological benefits have taught us is that side effects will always be there. Indeed, whether a patient's response to a drug will be considered a "side effect" or a "therapeutic effect" is purely a matter of our perspective and goals, and what is a side effect in one situation can be the therapeutic goal in another; think of retarded ejaculation with SSRIs, which can be a treatment for premature ejaculation; or of emotional numbing with dopamine blockers, which can be the source of its anti-manic effects. All effects are equally results of our intervening in the complex neurophysiology of the brain, and will remain with us, be it as therapeutic salve or iatrogenic thorn.

Inherent to the brain's complexity is the difficulty of ever achieving a one-to-one mapping of chemical manipulation to symptom (and of course to DSM diagnoses, which are not coherent enough to produce any consistent neurophysiological or genetic finding), so that amongst the myriad effects that medications produce will always be those which are undesirable. As I note in the introduction to this book, back in the 1990s, the enthusiasm for second-generation antipsychotics was premature, for we did not yet know the side-effects which would be discovered. So why should we be more optimistic about yet undiscovered, barely imagined neurochemical manipulations? Unwanted, harmful effects will always remain a part of the package.

Most likely, the expectation for more benefits in the future will be disappointed for another reason as well. Much of what we treat is the result, partly or mostly, of all sorts of psychosocial problems. For example disorders such as non-melancholic depression, anxiety and post-traumatic syndromes. I expect that for these problems, the psychopharmacological intervention will never be more than symptomatic.

Finally, the problems inherent to psychopharmacology such as passivity, stigma, a confusion between brain and psyche will not be solved by finding new neurophysiological targets. Indwelling catheters will never be an acceptable option for more than a small fraction of our most severe patients. We have indeed picked low-lying fruit, and it has been instructive about the pitfalls we can expect to dog us in psychopharmacology in the future as well. The reader must be cognizant of the likelihood that our psychopharmacological interventions will rarely or never provide on their own a solution for psychiatric misery.

One possible a new direction is the therapeutic potential of the rediscovered psychedelics. While some results have been promising [5], the jury is still deliberating. What is most original about this treatment is that it incorporates a psychotherapeutic intervention intended to guide the experience of the person receiving the drug. That means that our understanding of the therapeutic effect of the medication will focus on the experience of the person on a "trip", and not on the neurophysiological alterations which unquestionably occur. We have here, then, a psychopharmacological manipulation which respects, examines, and relies upon the subjectivity of the person. Perhaps the younger readers of this book will discover the where and when this can help.

References

1. Wegener G, Rujescu D. The current development of CNS drug research. Int J Neuropsychopharmacol. 2013;16(7):1687–93.
2. Prinz F, Schlange T, Asadullah K. Believe it or not: how much can we rely on published data on potential drug targets? Nat Rev Drug Discov. 2011;10(9):712.
3. George MS, Belmaker R. Transcranial magnetic stimulation in neuropsychiatry. Washington DC: APA Press; 2000.
4. Belmaker RH, Agam G. Deep brain drug delivery. Brain Stimul. 2013;6(3):455–6.
5. Goodwin GM, Aaronson ST, Alvarez O, Arden PC, Baker A, Bennett JC, et al. Single-dose psilocybin for a treatment-resistant episode of major depression. N Engl J Med. 2022;387(18):1637–48.

Chapter 23
Learning and Teaching Psychopharmacology: A resident's Point of View

Alexander Moshe Clayman

I am a psychiatrist-in-training. My journey through psychiatry has taken conventional and less conventional paths, all of which I have found value within. I became an MD in 2017 after graduating from the University of Malta, read for an M.Sc. in Public Health, finished my 2-year internship and worked a year in a bustling Emergency Department in Malta's university teaching hospital. Having worked my first rotation as a 23-year-old doctor in Malta's Victorian-era psychiatric institution (Mount Carmel Hospital), I explored other options, working in community mental health away from my home country, before returning to a recognized psychiatry training program.

When you are a medical student and later a junior doctor working in mental health, it is easy to become convinced that the only tools at your disposal are medications. In busy emergency departments and overflowing psychiatric wards, everybody is searching for the swiftest, most efficient way to alleviate suffering. Journeys of psychoanalytical discovery, open dialogues with family members or significant others and other such time-consuming and human-resource intensive activities are simply seen as less practicable solutions than a prescription. This kind of thinking is not entirely delusional: when you are a single psychiatric trainee on a night shift covering a general hospital and a regional emergency department it is truly unlikely that there will be time or space to explore and treat psychiatric phenomena in an ideal manner. It remains true that we must all learn how to use psychiatric medications competently.

As psychiatric trainees, we judge our teachers just as they evaluate us. We learn to recognise their idiosyncratic prescribing habits, we compare them to each other, to what we are taught by textbooks and by specialty board examinations. "The new guy in Ward 1 is on sulpiride... no points for guessing which professor they were assigned"... The truth is that individual psychiatrists seem to prescribe as if they had studied separate materials. I used to wonder to myself, "What does that professor know about sulpiride that evades all other psychiatrists in this department?"

The original version of this chapter has been revised. The correction to this chapter can be found at https://doi.org/10.1007/978-3-031-40371-2_24

R. H. Belmaker, P. Lichtenberg, *Psychopharmacology Reconsidered*, https://doi.org/10.1007/978-3-031-40371-2_23

Compared to other textbooks, this one is somewhat unorthodox in its approach, but it demonstrably draws from the same research that everybody else has access to. It presents empirical explanations for the positions it takes, and ultimately retains many of the core concepts familiar to students and practitioners of psychiatry: that antidopaminergics have some dampening effect on manic/psychotic symptoms, that lithium seems to balance mood and reduce suicide, and that antidepressants are useful in some people suffering severe depressive symptoms. There are some caveats. Benzodiazepines make it through the book surprisingly unscathed (compared to other commonly prescribed psychiatric medications) whereas stimulants such as methylphenidate (and other medications for behavioral disorders starting in childhood) come out the other end of the book licking their wounds, in need of intensive rehabilitation.

Reviewing this book as it was being written was fascinating and enlightening. I am equally fascinated (and grateful) that I was given the chance to share my thoughts about it. I have attempted to present my comments as a discussion, rather than a repetition of previous chapters. For further details, please refer to the chapters alluded to and their sources respectively.

Antidepressants and Sexual Psychopharmacology

Trainees should benefit from these chapters. The content does not present a radical departure from mainstream thought on treatment of depression, but does uniquely articulate important approaches to choosing antidepressant treatments: identifying subtypes of depression, tailoring side-effects, and clever use of biofeedback to improve psychiatric symptoms. Reference to the history of drug-development is welcome, and reference to key trials such as STAR-D should not be new to most psychiatric trainees.

Depression is now a broad concept, and people awarded the diagnosis of depression differ qualitatively and quantitatively. Depression is not a homogenous concept; a person diagnosed with depression can significantly differ from the next person with depression in terms of symptoms (sleep, appetite, energy levels, libido), etiology (reactive to a stressful life event, "out of the blue", relating to a medical condition, etc.) and prognosis. It makes sense to distinguish subtypes of depression because optimal treatments differ.

The discussion of side effects should interest trainees, especially those inclined to research. When SSRIs were essentially declared free of side-effects compared to TCAs, this was because developers did not know which side-effects they should have been looking for. It surprised me that the scientific community got caught out so easily in this regard, running with the premature declaration that SSRIs have fewer side-effects than TCAs: the reality that trainees should re-orient themselves with is that the side-effect profiles are simply different.

This being said, the border between an effect and a side-effect is dependent on perspective: the atypical antidepressant mirtazapine commonly increases appetite.

This is a side-effect for some, but the primary intention of its use in cachectic cancer patients in palliative care settings. As we familiarize ourselves with the overall effects of medications available in our local formulary, we become more competent prescribers. Different people will have different boundaries for what they welcome or tolerate as side effects, and as clinicians we need to be prepared to discuss this knowledgeably and openly with our patients.

I was impressed by the suggestion to use phosphodiesterase-5 (PDE-5) inhibitors like sildenafil to treat depression in patients with sexual dysfunction. If sexual dysfunction (e.g. impotence) is conceivably both a causative factor and/or symptom of depressed mood and low self-esteem, it may make sense to prescribe a medication that primarily acts upon changing the bodily symptom. Surely this section is showing that psychopharmacology can take a holistic approach: it recognizes the interaction between brain, mind and body and suggests that psychiatrists too can intervene at the level of the body (specifically by modulating genital vasculature) to improve mental wellbeing. As is so often the unfortunate case in medicine, the treatment options for women are dwarfed by the treatment options for men; as reflected in this book by the length of the discussion on men's versus women's issues.

Antipsychotics

This is a meaty chapter which focuses much energy on understanding how antipsychotics were developed, their theorized mechanism(s) of action and how best we ought to use them in the future. The dopamine hypothesis of psychosis is well-described but like many coherent narratives, it is exposed to have not told the whole story. Psychosis is complex, and though a theory pinning it on just one aberrant neurotransmitter has been helpful in exploring therapeutic avenues, it should be taken with a pinch of salt. This equally applies to the idea that dopamine-receptor blockers (the more descriptive name for antipsychotics) specifically target a pathological process underlying psychosis.

The discussion of pharmaceutical industry tactics (e.g. the non-disclosure of metabolic side effects of certain second generation antipsychotics) is at times chilling, but necessary reading for trainees. As we welcome new drugs to the market, we should caution ourselves to be aware of the future unknowns before becoming too enthusiastic in our prescribing habits.

Referencing the CATIE study should not be news to trainees since it is (rightly) a favorite topic in specialty board examinations. In a practical sense, the authors mirror CATIE's findings with regards to the choices between antipsychotics: "they all kinda' work the same". Just as we explored with antidepressants, efficacy remains stable but side effects differ considerably. It is once again our duty to be deeply familiar with the effects of the drugs we prescribe and to foster, wherever practicable, our patients' understanding and cooperation.

The conclusion that a DSM diagnosis of schizophrenia (or one of the other psychoses) is not an automatic indication for long-term or life-long antipsychotic treatment is quite a drastic departure from mainstream psychiatry. It was arrived upon by recalibrating the balance between risk and benefit: antipsychotics have serious, not trivial side-effects and, coupled with a moderate efficacy (the fact that some people recover without treatment) there are ethical implications to prescribing them life-long. Over the course of my training I have heard these arguments being made, but they would remain the "incorrect" answer in a specialty board examination in most places.

Mood Stabilisers

Once again, to understand the present, it helps to know the past. Readers are taken on an omnibus ride through the history of lithium and it isn't a ride I'd want to miss. That lithium is natural, cheap, unpatentable and that much key research about it was possibly deliberately undermined by commercial influence in FDA policy suggests that we should maintain our sights on this peculiarly simple inorganic compound. In spite of this, trainees may be surprised to read that lithium is neither "gold standard" nor "specific" for any disorder.

Throughout most of the commentary about lithium, the anti-epileptic mood stabilizers, and the dopamine-blocking mood stabilizers, the message is that a lot of these medications work, they're mostly equivalent (i.e. there is no clear superior compound) but they are not the same. Patient history and previous response to a specific compound is recommended over simply following guideline recommendations. This makes sense when we are reminded that the randomized control trials which feed meta-analyses which feed guideline recommendations are based upon a sample mean, whereas psychiatrists (and trainees) are interested in individuals.

The most striking hypothesis to be made during this chapter was the idea that clozapine and lithium, being the two psychiatric medications known to significantly decrease suicide, are also two drugs that demand serious laboratory follow-up, and that perhaps regular contact with medical services is responsible for their anti-suicidal effect. Psychiatrists and trainees often gossip about medications in doctors' offices and in between ward rounds (the osmotic educational effect of being a doctor in a teaching hospital, perhaps) so I was surprised to hear a completely new and truly tantalizing hypothesis. Of course, it remains speculative until proven otherwise - possibly only the most motivated people agree to the hassle of weekly CBCs/lithium level - but to me this seems like an exciting new avenue for researchers.

As a trainee studying guidelines and board examinations, we know that it is not only clozapine and lithium that should be monitored with laboratory tests. Guidelines differ slightly, so I will avoid quoting specific time-frames, but people taking atypical antipsychotics should have their BMI, ECG and lipid profile measured regularly, people taking certain mood stabilisers and antidepressants should have their liver enzymes and electrolytes measured… If by inviting these people in for regular tests we create chances to engage with them and detect psychological distress, perhaps this is an important and beneficial side effect of psychopharmacology.

OCD

The only surprising facet of the OCD chapter was how well it aligns with what I have learnt from other textbooks and guidelines. As part of my CBT-training during my child and adolescent psychiatry rotation, I was assigned a young person with disabling OCD, so this is a topic I know relatively well. If only all of psychiatry were this straightforward: sticking to SSRIs and serotonergic TCAs (i.e. clomipramine) coupled with the "superficial" therapies (CBT and exposure and response prevention) without delving too deep into the root psychological issues seems to be the winning recipe.

Though no definitive organic cause has been identified, the existence of PANDAs and the fact that deep brain stimulation and psychosurgery have some role in OCD treatment suggest that we may not have reached the end of the road in our research into the etiology of this condition.

Anxiety

A wonderful feature of this textbook is that it does not ignore historical contexts and developments. I find it extremely helpful to hear about how psychiatrists of the past used to practice and I benefit from understanding which medications used to be available. Modern training programs in psychiatry do not usually require practical knowledge about barbiturates or meprobamate (in fact, I can barely spell barbiturates), but we are extensively taught about benzodiazepines. Having some insight into the history of anti-anxiolytics, I can appreciate that benzodiazepines are an improvement over barbiturates in terms of their side effects and addiction potential, but I still struggle to come to terms with how positively they are treated in this chapter.

Compelling arguments for the use of benzodiazepines exist, especially if there is (as suggested) frequent communication/collaboration between physician and patient in adjusting the dose, and eliminating the drug once its acute effects are no longer beneficial. Benzodiazepines are commonly-used medications in specialties other than psychiatry so it is natural that many doctors will feel comfortable prescribing them - they are clearly not going to go away any time soon.

Nevertheless, in my relatively short medical career, I have already come across hundreds of patients who have been taking daily benzodiazepines since before Elton John was famous with no discernible benefit with the exception of the initial period of treatment. In one rural community I worked in, illiterate elderly patients would know nothing about their medical or medication history, with the exception of the name and dose of their benzodiazepine treatment. There are legitimate concerns about benzodiazepines regarding their paradoxical effects (disorientation and aggressive behavior), their effects on cognition, issues around addiction/tolerance as well as their contribution to death by overdose which should not be brushed aside so lightly [1, 2].

Monoamine reuptake inhibitors are not given much prominence in this chapter, except for their relevance in panic disorder. Textbooks and other sources of conventional knowledge actually give similar recommendations, but from a different direction: they recommend SSRIs for most forms of anxiety, but specifically recommend against benzodiazepines in panic disorder.

Stimulants and Children

The basis of ADHD as a concept is questioned here - this would not be the first time psychiatric academia has been skeptical of ADHD. The basis of the ADHD diagnosis in this chapter is framed as hyperactivity, whereas DSM/ICD puts more emphasis on attention deficit. Medical literature has been critical of geographical overdiagnosis of ADHD to the exclusion of other determinants of "deviant" behavior such as childhood trauma, abuse, attachment disorders and other environmental factors [3].

Genuine questions are raised about the paradoxical effects of stimulants on the human brain: is there a difference between stimulant effects on a pre- VS post-pubertal brain? How can we treat hyperactive behavior with stimulants sometimes and at other times with an antipsychotic like risperidone?

Not mentioned in this chapter are the adults who often present to forensic and addiction psychiatry services for harmful cocaine use. On deeper questioning, many of these adults will admit to using cocaine not to get high or to experience a "buzz" but to calm down and achieve a state of relaxation. Is this part of the "paradox" of stimulant effects on the brain (stimulation/mania/psychosis VS relaxation/focus) or does it suggest that there are individuals who respond differently to stimulant medications?

I agree that all trainees would do well to be aware of cultural and educational pressures when considering an ADHD diagnosis and prescription of stimulant medications. Does little Tony really have a short-attention span or are his classes simply boring? That attention and activity fall on a spectrum is relevant, but is not in itself a reason to avoid prescribing stimulants or diagnosing ADHD: blood pressure and glucose levels are also found on a spectrum and few would argue against establishing thresholds for pharmacological intervention when these thresholds are breached.

Clinical Trials, Diagnosis, Placebo and the Future

As someone with an analytical personality and a nerdy obsession with public health, I derived most pleasure from reading these chapters. In terms of getting to grips with the principles (and problems) of psychopharmacology, these are the most important chapters in the book.

The behavior of pharmaceutical companies with regards to SSRIs in the 1990s is the subject of much discussion, popularized at least in Europe by Ben Goldacre's extensive enquiry into publication bias and other tricks [4]. Pharmaceutical company trickery led the UK's Royal College of Psychiatrists to introduce extensive "evidence-based practice" content into their membership exam's syllabus. When justifying the addition of the quite rigorous epidemiology and biostatistics modules to their examinations, members of the RCP's Critical Review Paper Panel wrote "The tendency for published research funded by the pharmaceutical industry to favour new therapies is well known... Perhaps one of the most famous examples of potentially inappropriate comparators was the use of high-dose conventional antipsychotics in randomised controlled trials of atypical antipsychotics for the treatment of schizophrenia." [5].

The chapter on clinical trials includes but goes beyond the familiar critique of medical research's entanglement in pharmaceutical company influence. It discusses core epidemiological concepts such as Bayesian statistics and the difference between statistical and clinical significance (so what if $P < 0.05$ if what we are measuring is clinically irrelevant?). It contrasts the power of the clinical trial with the art of medicine's supposed focus on the individual. In medical school, we learn physiological principles based upon the assumption that we are discussing a 75 kg male specimen. Clinical trials are similar in that they often limit participation to people with a single diagnosis, aged 18–65 who do not take other medications. This drive to exclude "noise" may prove to be misguided since it decreases generalizability and validity. In the future, trainees should be on the lookout for Real World Evidence and pragmatic clinical trials, or at least clinical trials which attempt to minimize inclusion/exclusion criteria to allow for greater generalizability (the STAR-D trial is an example of the latter). I was heartened to read the comments praising the value of and encouraging small-scale research. There should also be a place for qualitative research in psychopharmacology: had psychiatry paid more attention to the lived experience of those receiving SSRI treatment, perhaps the complex "personality-blunting" side effects would have been recognized sooner than they were.

Discussions about the validity of psychiatric diagnoses are invariably present in the main textbooks I have studied from, but they are rarely the subject of post-graduate examinations. Perhaps this is because nosology in psychiatry is a dangerous subject which causes our specialty to feel insecure: we cannot prove our diagnoses under a microscope or in a CT scanner, and our most concrete diagnoses drift toward other specialties (tertiary syphilis is an infectious disease, Parkinson's a neurological disorder, etc). At least in my experience, there is a healthy awareness of the fragility of psychiatry's diagnostic systems among trainees and psychiatrists, even if healthcare systems insist on cooperation with DSM/ICD. In the first psychiatric hospital I worked at, all discharge summaries were required by the hospital records office to have an ICD-10 diagnosis and code. Some doctors, including veteran psychiatrists, would refuse to humor these requirements, much to ire of the bureaucrats in the records office. The intern writing the discharge summary would often ask the senior doctors "what was Mr X's diagnosis?" as he was being discharged. In other medical specialties the diagnosis is not routinely obscured until the patient's last day at hospital.

What could be more fascinating and omnipresent in all of medicine, than the placebo? We tend to take placebos for granted, but this may be the only way they can be allowed to function. Once we know how a magic trick works, does it not lose its magic? Perhaps not necessarily. In addition to the "true placebo" effect (the conscious expectation of improvement aroused by treatment and the conditioned, unconscious response to the act of receiving treatment), other determinants of improvement in a patient's condition should remain in the back of any treating doctor's mind. We would do well to recall Voltaire's wise remark,. "the art of medicine is to amuse the patient while nature cures".

In all of my medical and psychiatric education and experience, the importance of a therapeutic relationship/alliance has been consistently impressed upon me. People like to see the same psychiatrist (or doctor, or nurse, or therapist, etc) and become disgruntled if they see a new face each time they visit the clinic. Combining the powers of a therapeutic relationship, the "top-down" approach of medications' placebo effects with the "bottom-up" receptor-altering pharmacological effects of medications should be what psychiatric trainees are learning to master.

As for the future of psychopharmacology, I haven't got a clue. I do not have any bright ideas about which receptor or neurotransmitter we should go after next. I do hope that researchers, clinicians and drug-developers can work together sensibly and without dishonesty. I hope that legal frameworks will allow for research to continue on substances that were previously illegal, (because I am excited about the possibility of combining psychotherapy with medications like psilocybin and LSD). I am optimistic that as we understand more about the nervous system, we will perhaps repurpose existing medications (like ketamine) and develop other biological therapies (like neurofeedback) to help alleviate mental suffering.

Concluding Remarks

Rather than offering specific prescribing advice, this book offers unique perspectives on key aspects of psychopharmacology. These key areas of discussion are as follows:

- The consideration of diagnostic categories (a tendency towards broadening of diagnostic labels) and thus non-specificity of diagnoses (e.g. subtypes of depression).

- The influence of diagnosis on treatment decisions - whether we should approach psychopharmacological decision-making in terms of syndrome (psychosis) rather than diagnosis (delusional disorder, schizophrenia). Of note, the old notion that lithium is specific to bipolar disorder is challenged here.
- Reopening value judgements such as the risk-benefit analysis of broadly-accepted psychopharmacological treatments: is it worth prescribing SSRIs so widely given their propensity to depress sexual function, coupled with their relatively mild efficacy?

I was disappointed to note the absence of a discussion about the ethics of involuntary treatments, specifically injectable antipsychotics. Forced treatment is practically (though not completely) limited to psychiatrists prescribing injectable dopamine receptor blockers, of which the effects (including side effects) are now known and are often deeply distressing to the recipient. Medications should be discussed within the contexts that they are used…if this discussion is not suited to this textbook, then I am unsure of where it would be relevant.

My final comment would be that: anybody remotely connected to the study or practice of modern psychiatry should read this book once they have been introduced to the existence of the commoner psychiatric medications. Among others, medical/nursing students, psychiatry/neurology trainees and pharmacy students would do well to absorb the spirit of this book. Compared to other textbooks in the field, there is not much to memorize from within these pages, but there is a lot to comprehend, and mindful comprehension is what makes or breaks humane, holistic, patient-centered professionals.

References

1. Mathieu C, et al. Patterns of benzodiazepine use and excess risk of all-cause mortality in the elderly: a Nationwide cohort study. Drug Saf. 2021;44:53–62.
2. Xu KY, Hartz SM, Borodovsky JT, Bierut LJ, Grucza RA. Association between benzodiazepine use with or without opioid use and all-cause mortality in the United States, 1999-2015. JAMA Netw Open. 2020;3:–e2028557.
3. Van der Kolk BA. The body keeps the score: brain, mind, and body in the healing of trauma. New York: New York, Penguin Books; 2015.
4. Goldacre B. Bad pharma. London: Fourth Estate; 2012.
5. Carney S, Warner J, Ahmad S, Rands G, Suleman S. Teaching, learning and assessing evidence-based psychiatry. Psychiatrist. 2011;35(5):192–5. https://doi.org/10.1192/pb.bp.110.030056.

Corrections to: Learning and Teaching Psychopharmacology: A resident's Point of View

Correction to:
Chapter 23 in: R. H. Belmaker, P. Lichtenberg,
Psychopharmacology Reconsidered,
https://doi.org/10.1007/978-3-031-40371-2_23

In the original version of the book, Dr. Alexander Moshe Clayman's name was omitted.

The author's name has been included in the corrected publication.

The updated version of this chapter can be found at https://doi.org/10.1007/978-3-031-40371-2_23

C1

R. H. Belmaker, P. Lichtenberg, *Psychopharmacology Reconsidered*,
https://doi.org/10.1007/978-3-031-40371-2_24

Index